THE BIOLOGY
OF TRANSCENDENCE

▼

A Blueprint of the Human Spirit

JOSEPH CHILTON PEARCE

Park Street Press
Rochester, Vermont

Park Street Press
One Park Street
Rochester, Vermont 05767
www.InnerTraditions.com

Park Street Press is a division of Inner Traditions International

The Library of Congress has cataloged the hardcover edition as follows:
Pearce, Joseph Chilton.
    The biology of transcendence : a blueprint of the human spirit / Joseph Chilton Pearce.
        p. cm.
Includes bibliographical references and index.
    ISBN 0-89281-990-1 (hardcover)
    1. Spiritual life—Psychology. 2. Brain—Religious aspects. 3. Evolution—Religious aspects. I. Title.
    BL624 .P42 2002
    128—dc21
                                                                2002002530
ISBN of paperback edition: ISBN 978-1-59477-016-6

Printed and bound in the United States at Lake Book Manufacturing, Inc.

10 9 8 7 6 5 4

Text design and layout by Priscilla Baker
This book was typeset in Caslon, with Trajan and Legacy Sans as display typefaces

In memory of Lucy Jane Whitehead

O lost and by the wind-grieved ghost
Come back again . . .

—THOMAS WOLFE

CONTENTS

Part Three

BEYOND ENCULTURATION

ACKNOWLEDGMENTS

With gratitude to: Lew Childre and staff of HeartMath Institute for their warm friendship, brilliant research, and training—I owe them much of the brain-heart information and most of the illustrations and diagrams used in this book; the Siddha Yoga Foundation and my teachers Muktananda and Gurumayi, for giving me an understanding of the heart as only great beings could, as well as years of rich experiences of the heart no science could discover nor books provide; Maria Colavito and Antonio de Nicolas for introducing me to their concept of bioculturalism; Michael Mendizza and his Touch the Future Foundation for long friendship and support; David Spillane for his stimulating mind and endless generosity with books, articles, clippings, and research reports. Thom Hartmann, for his generous help and suggestions, for introducing me to Bear and Company, and for bringing to us Robert Wolff's *Original Wisdom* (see Epilogue); Keith Buzzell for sharing his insight and understanding, and for his concern for children's welfare; Bruce Lipton for his brilliance and knowledge, high-spirited generosity, endless energy on our behalf, and for being fun to work with; James P. Carse for his wonderful books that unhinged my comfortable notions time and again; Gil Bailie, for his book *Violence Unveiled,* which marked a turning point in this book of mine; Cheryl Canfield for her book *Profound Healing,* and for helping me personally to come to terms with the toll-taking of Father Time; Allan Schore for his monumental work, *Affect Regulation and the Origin of Self,* from which I have liberally stolen; George Jaidar for his insight into culture and enculturation; Charles Sides and Gregg Korbon for their patient readings and constructive criticisms of several early drafts of this book; Matthew Fox, not only for his books on Eckhart and the Beguines, but especially for *Original Blessing,* a great testament to the human spirit. Finally, a very special thanks to Elaine Sanborn for her prodigious editing efforts on behalf of this book, a task that would have daunted lesser souls.

INTRODUCTION

Looking up at the starry sky, poet Walt Whitman asked:

When we become the enfolders of those orbs,
and the pleasures and knowledge of everything in them,
shall we be satisfied then?
And my spirit answered No, we but level that lift
to pass and continue beyond.

"The ability to rise and go beyond" is the definition of *transcendence* and the subject explored in the following pages. While this force constitutes our nature and fires our spirit, an honest exploration of it must contend with this counterquestion: Why, with a history so rich in noble ideals and lofty philosophies that reach for the transcendent, do we exhibit such abominable behaviors? Our violence toward ourselves and the planet is an issue that overshadows and makes a mockery of all our high aspirations.

Sat Prem, a French writer transplanted to India following World War II, recently asked this question: "Why, after thousands of years of meditation, has human nature not changed one iota?" In the same vein, this book asks why, after two thousand years of Bible quoting, proselytizing, praying, hymn singing, cathedral building, witch burning, and missionizing has civilization grown more violent and efficient in mass murder? In exploring the issue of transcendence, we explore by default the issue of our violence. The two are intertwined.

A great being appeared some two millennia ago, looked at our religious institutions with their hierarchies of power, professional classes, policy makers, lawyers, and armies, and observed that we should "know them by their fruits." That is, we should ask: What are the actual, tangible results of these lofty religious institutions that we have known throughout history? If we examine them by the fruits they produce, rather than by the creeds, slogans, concepts, and public relations that sustain them, we would see that spiritual transcendence and religion have little in common. In fact, if we look closely, we can see that these two have been the fundamental antagonists in our history, splitting our mind into warring camps.

Neither our violence nor our transcendence is a moral or ethical matter of religion, but rather an issue of biology. We actually contain a built-in ability to rise above restriction, incapacity, or limitation and, as a result of this ability, possess a vital adaptive spirit that we have not yet fully accessed. While this ability can lead us to transcendence, paradoxically it can lead also to violence; our longing for transcendence arises from our intuitive sensing of this adaptive potential and our violence arises from our failure to develop it.

Historically our transcendence has been sidetracked—or derailed altogether—by our *projection* of these transcendent potentials rather than our development of them. We project when we intuitively recognize a possibility or tendency within ourselves but perceive this as a manifestation or capacity of some person, force, or being *outside of ourselves*. We seem invariably to project onto each other our negative tendencies (". . . if it weren't for the likes of you . . . that government . . . those people . . ."), while we project our transcendent potentials onto principalities and powers "out there" on cloud nine or onto equally nebulous scientific laws. The transcendence we long for, then, seems the property of forces to which we are subject. Like radar, our projections bounce back on us as powers we must try to placate or with which we must struggle. Perennially our pleas to cloud nine go unheeded, our struggles against principalities and powers are in vain, and we wander in a self-made hall of mirrors, overwhelmed by inaccessible reflections of our own mind. Handed down through millennia, our mythical and religious projections take on a life of their own as the cultural counterfeits of transcendence.

Culture has been defined by anthropologists as a collection of learned survival strategies passed on to our young through teaching and modeling. The following chapters will explore how culture as a body of learned survival strategies shapes our biology and how biology in turn shapes culture. Religious institutions, cloaked as survival strategies for our minds or souls, are the pseudo-sacred handmaidens of culture brought about through our projections of the transcendent aspects of our nature. Thus this trinity of myth, religion, and culture is both the cause and source of our projections.

Each element of the trinity brings the other into being and all three inter-locking phenomena—myth, religion, and culture—are sustained by the violence they generate within us.

Our greatest fear, the late philosopher Suzanne Langer said, is of a "collapse into chaos should our ideation fail us." Culture, as the collected embodiment of our survival ideation, is the mental environment to which we must adapt, the state of mind with which we identify. The nature or character of a culture is colored by the myths and religions that arise within it, and abandoning one myth or religion to embrace another has no effect on culture because it both produces and is produced by these elements.

Science has supposedly supplanted religion—but it has simply become our new religious form and an even more powerful cultural support.

If our current body of knowledge, scientific or religious, is threatened, so are our personal identities, because we are shaped by that body of knowledge. Such threat can lead us to behaviors that run counter to survival. This book explores how our violence arises from our failure to transcend, and how our transcendence is blocked by our violence; how it is that culture is a circular stalemate, a kind of mocking tautology, self-generative and near inviolate. That we are shaped by the culture we create makes it difficult to see that our culture is what must be transcended, which means we must rise above our notions and techniques of survival itself, if we are to survive. Thus the paradox that only as we lose our life do we find it.

A new breed of biologists and neuroscientists have revealed why we behave in so paradoxical a manner that we continually say one thing, feel something else, and act from an impulse different from either of these. After centuries of bad remedies prescribed for a disease that has been wrongly diagnosed, this new research gives us the chance to remove the blocks to the transcendent within us and allows us to develop a nature that lies beyond rage and violence.

A major clue to our conflict is the discovery by these new scientists that we have five different neural structures, or brains, within us. These five systems, four of them housed in our head, represent the whole evolution of life preceding us: reptilian, old mammalian, and human. Nature never abandons a good idea but instead builds new structures upon it; apparently each new neural structure we have inherited evolved to correct shortcomings in or problems brought about by nature's former achievements. Each neural creation opened life to vast new realms of possibility and, at the same time, brought new problems, thus calling once again for "rising and going beyond" through the creation of yet another neural structure. Thus, while we refer to transcendence in rather mystical, ethereal terms, to the intelligence of life, transcendence may be simply the next intelligent move to make.

As long intuited by poet and saint, the fifth brain in our system lies not in our head, but in our heart, a hard biological fact (to give the devil of science his due) that was unavailable to the prescientific world. Neurocardiology, a new field of medical research, has discovered in our heart a major brain center that functions in dynamic with the fourfold brain in our head. Outside our conscious awareness, this heart-head dynamic reflects, determines, and affects the very nature of our resulting awareness even as it is, in turn, profoundly affected.

Within this mutually interdependent system lies the key to transcendence and the resolution of our perennial and now near-terminal tendency toward violence. We can, through considering this new research, become more aware of and cooperate with nature's head-heart dynamic, the dynamic of intelligence and intellect, of biology and spirit.

As used here, spirit is that unknown power impelling us to rise and go beyond. Poet Dylan Thomas defines it as:

The force that through the green fuse drives the flower
Drives my green age . . .

The intelligence of the heart brain embodies this elusive driving force, a fact we can grasp if we distinguish between intelligence and intellect as we must between the spiritual and the religious. In an efficient biological unfolding, the intelligence of our heart and the intellect in our head should function as an interdependent dynamic, each influencing and giving rise to the other. The breakdown or impairment of this reciprocal action is brought about by its cultural counterfeits of myth and religion. This, in turn, brings about both our fundamental split of self and our self-wrought woes—providing an explanation for why it is that we build bombs with one hand even as we gesture toward peace and love with the other.

Two geniuses of the late thirteenth and early fourteenth centuries, the Dominican monk Meister Eckhart and the Spanish Sufi philosopher Ibn Arabi, spoke of "creator and created giving rise to each other." This is an equally accurate if more arcane way of looking at the relationship between intelligence and intellect, each of which is designed by evolution to give rise to the other. The proposal, of a new wave of biologists in our own time, that "mind and nature are one" is but another recognition of this dynamic, and the recently discovered heart-brain reciprocation clearly demonstrates the actual means by which this "dual birth" takes place—or should.

Ibn Arabi and Meister Eckhart claim that we are an integral part of this dynamic, indissolubly one with it rather than a victim of the process. Their predecessor Jesus pointed out the same transcendent fact and got strung up for his trouble. Such insights regarding the creative dynamic within us have generally resulted in whoever proffered them being led to the stake or block, but have seldom fallen into the public domain. Notions like this are heretical to the reigning mind-set or power structure of any age and are generally mistranslated or eradicated.

That creator and created give rise to each other is the major principle on which this book is based. This dynamic is stochastic, however (*stochasm* is a Greek word for a system that is random but purposeful); accident and chance underlie every facet of our life, much as we would like to eradicate them—but to eradicate stochasm would turn life into a mere mechanism, which it is not.

From this background I make two proposals here that are necessarily hypothetical: First, the crux of our ever-present crisis hinges on failure to develop and employ both the fourth and newest brain in our head (one added quite recently in evolutionary history) and its dynamic interactions with our heart brain. Second, the great saints and spiritual giants of history (even though overlaid with myth and fantasy by cultural counterfeits) point toward, represent, or manifest for us our next evolutionary step, a transcendent event that nature has been trying to unfold for millennia.

Creator and created as a co-inspiring dynamic make imperative a simple natural law: Intelligence, no matter how innate or genetically encoded, can unfold within us only when an actual model for that intelligence is given us. All dynamics must have their generative source, even if the source can never be factually determined—if there are two mirrors reflecting each other in an infinite regress, which one could we say initiates the reflection? From the beginning of our life, the characteristics of each new possibility must be demonstrated for us by someone, some thing, or an event in our immediate environment—but the same chicken-egg paradox will always emerge if we try to determine or bring closure to the riddle of an origin.

This need for a model is acutely the case with a new and unknown form of intelligence such as that offered by our fourth brain and heart brain. The striking contrast between our ordinary human behavior and the actions of the great beings of our history (Jesus, Krishna, Lao-tzu, Buddha, Eckhart, George Fox, Peace Pilgrim, and a long line of like geniuses) is what makes these figures stand out in time even as shifting or warping history itself. Our great beings arise through a natural process that we will explore here, though the process unfolds in that infinite regress that obscures its origin. They come into being as models of nature's new possibility, our next evolutionary step manifested by our newest neural structure, transcending violence to create a new, viable reality.

In every case, however, rather than developing the capacities these great models of history have demonstrated, humankind has projected both the capacities and the image of the models demonstrating them. That is, we invariably build religions around our spiritual giants or use them to support a religion in order to avoid the radical shift of mind and disruption of culture these rare people bring about, shifts we interpret, ironically, as threats to our survival and thus instinctively reject. Biocultural effects, once initiated, tend to self-generate. Projected by us, we perceive the behaviors demonstrated by our great models as powers out there to which we are subject, rather than as potentials within ourselves to be lived.

Our fourth brain is the way by which the intelligence of our heart can guide the intellect in our head from its ancient survival strategies to a new and greater form of intelligence. But nature's dilemma—and thus ours as we are, in effect, nature herself—has been how to stabilize a new and largely undefined intelligence in a powerful neural environment millions of years old. Though nature has provided appropriate models as the opportunities have arisen, behaviors encoded in our ancient primary brains are thoroughly entrenched, whereas the new ones offered are tentative at best. And it is from just this tenuous uncertainty of a higher intelligence locked into our firmly entrenched survival systems that our wild contrasts of lofty ideals and deadly real behaviors emerge.

The following exploration revolves around the insights gleaned from the research of this new cadre of biologists and neuroscientists and from the ideals and behaviors modeled for us by the great beings of our history—specifically by, in my opinion, the greatest model of all, Jesus. An odd couple to find between the covers of the same book, you might think—Jesus and the new biologist. But if we drop the mythical and/or religious projections surrounding Jesus, we will discover a common ground.

No matter that we might personally reject religion and myth, the survival culture that both spawns and is spawned by these two is still very much with us, converting all our efforts, scientific or spiritual, to its service, and keeping us locked in our primitive survival modes of mind. As model of a new evolutionary intelligence, Jesus met and continually meets a grim fate at the hands of this cultural effect. But the cross, the instrument of his execution, symbolizes both death and transcendence for us—our death to culture and our transcendence beyond it. If we lift the symbol of the cross from its mythical shroud of state-religion and biblical fairy tale—which is to say, if we can rescue Jesus from the Christians—then the cross proves to be the "crack" in our cultural cosmic egg.

It is toward this crack that this book points, as did my first book half a century ago. May this new one throw more light and help us to open ourselves to nature's new mind, wherein lies our true survival.

PART ONE

NATURE'S TRANSCENDENT BIOLOGY

Some Organic Details

▼

A SAGA OF UNCONFLICTED BEHAVIOR

I n my twenty-second year, World War II and the Army Air Corps behind me, I had three "blackout" experiences that ushered me into the world of subtle or psychic phenomena. All three took place within the same month, concerned the same event, and followed the same pattern—and they upset my roommate, who happened to be there as witness during each occasion. At the start of each experience, an enormous weight would suddenly bear down on me, literally pressing me out of my ordinary conscious state. The first time, it occurred as I crossed the room, and I dropped to the floor like a stone. I then found myself in a state of clear if bodiless awareness, observing the hand of my girl, the single greatest love of my life, who was some three hundred miles away, writing me a letter to explain why our relationship of four years must end. Three different times she wrote, explaining her case in different ways, and each time some corresponding knowing within me knocked me out of my body to observe her in the very moment of her writing. Each time, on returning to my usual state, I went into a most unusual emotional tailspin of no small proportion, my physical heart gripped in anguish, my roommate aghast and perplexed at my behavior.

Each time, the actual letter arrived a few days after in the mail my roommate brought in. Without taking the letter from him I quoted the exact contents of her correspondence from the copy that had been burned into my brain during my previous "vision." Each time he opened the envelope, read its contents, and looked puzzled—my report was identical to the missive.

Such events could be explained as simple precognizance, remote viewing, or some similar parapsychological phenomenon, except for one critical point: In that peculiar, subtle world of consciousness that I entered those three times, I was directly present with her being itself, the very core and spirit of her. I was not simply in her presence but was somehow fused with her presence—and being one with her in this way was the most unusual and unmistakable state I had ever known. In that direct presence I argued passionately with her concerning her decision, which was a veritable death sentence to me. And she spoke with me in her patient, gentle manner, explaining her case. We were each discrete and separate from our bodies, mine knocked out, hers busy writing, yet both of us together formed a peculiar unity of communion, observing her hand writing that fateful letter.

Later, when I read Carl Jung's theory of the anima, I felt that Jung had but a small angle on this intense and magnificent mystery. I had experienced my living anima on a level I had not known in the flesh itself. Years later, this subtle, ethereal world just beyond the material one proved to be the gateway to the most intense mystical experience of my life, an event of such magnitude that it nearly ruined me for the ordinary world thereafter.

Among many things, this fortieth-year event gave me to know that human sexuality, when it unfolds within a spiritual mantle of love, is a gateway to the highest transcendence. The earlier form of this experience—the "blackouts" when I was twenty-two—led eventually to a bizarre, nonordinary state for which I borrowed, or stole, the scholarly sounding term *unconflicted behavior*. This was an ongoing series of episodes that stretched over my twenty-third year and laid the foundation for my first book—*The Crack in the Cosmic Egg*. Although in that book I gave no detailed account of the major crack in my own egg, which this unconflicted behavior brought, I did approach the subject obliquely out of concern for my credibility in the eyes of my readers. (I began that book in 1958, which was a most conservative period as contrasted with 1970, the year I sold the book, when the New Age era had already burst upon us.)

The origins of this phenomenon of unconflicted behavior lay in my conviction that a major part of me had died on the loss of my anima-love the year before. The experience arose from of a kind of pseudo-suicidal recklessness that seized me, an "I couldn't care less" disregard of consequence that bordered on the irrational. Pushing this reckless abandon to extremes led to a breakthrough of knowing that took place within me with no transition or preparation. I discovered how to bypass my body's most ancient instincts of self-preservation, which resulted in a temporary absence of all fear and subsequent abandonment of all caution. This enabled me, at particular times, to accomplish things that would have been considered impossible under the ordinary conditions of our world.

In *The Crack in the Cosmic Egg* I recounted how, at a gathering of dormitory mates around a table, I demonstrated that fire didn't have to burn me. We all smoked back then and I used up a full pack of Pall Mall cigarettes (long unfiltered furnaces) to demonstrate my assertion. I puffed to maximum and then held the glowing ends of the cigarettes against my hands, fingers, wrists, then face and eyelids, grinding the tips into my skin. I concluded with getting three going full blast, then holding the lighted ends between my lips and blowing sparks about the table. During all of this I experienced intensity of feeling but no pain and had absolutely no trace of trauma on my skin the next day. As I pressed the cigarette to my skin each time, I knew with complete certainty that there would be no damage, and none occurred. This led a couple of physics majors in the group to test a cigarette tip's temperature, which proved to be 1,380 degrees Fahrenheit— only a bit more than half the temperature of a genuine fire walk, but hot enough to impress my fellow students.

This sort of unconflicted behavior manifested, it seemed, from a split-second recognition, without qualification or rationale, that death was a foregone conclusion, an integral part of that very event, that death was already within me. Death was not a possibility to be avoided but a fact to be accepted as it was already accomplished—death had already happened. I was struck by the hilarity of the thought "You can't kill a man twice," and would find myself in a state of ringing clarity I thought of as a world of invisible taut brass wires, though I have no notion where that image came from.

Having accepted death without hidden qualification, it was clear to me that I could not be threatened by the possibility of death or harm. During each incident I felt oddly invulnerable—and *was*, at that particular time. I seemed to stand on the cusp of being and nonbeing, to walk the line between subtle and physical, observing but not fully occupying my body. This shift of perspective gave what the anthropologist Mircea Eliade termed the ability to "intervene in the ontological constructs of the universe." This was Eliade's scholarly description of the nonordinary events brought about by Tibetan yogis, with whom he spent a decade in the 1940s. Years later I read his account, *Yoga: Immortality and Freedom* (New York: Pantheon, 1958).

I found that in any happening, through a kind of willful and voluntary throwing away of self-preservation, the ordinary course of events could be reversed, changed, or modified. This was not one part of my mind playing games of "let's pretend" with other parts, nor some lofty psychological or spiritual death of ego or loss of self. This was a genuine acceptance of death as a certain part of that moment, of knowing I held my nonbeing within my being. Therefore, there was nothing to lose! I found that in this state not only did fire not have to burn me, but also gravity did not have to hold me in the safety of its usual grip and cause did not have to produce its usual effect.

To find that the structure of reality was negotiable when I was free of all internal conflict was a momentous discovery for me—as was my realization that all internal conflict is produced by our fear of possible harm or death. The irony of this is that there exists for us a state in which harm really can't occur within the confines of a particular single event if we bypass our block of fear and open to this other perspective.

No matter how many times I experienced unconflicted behavior, however, my usual fear of death or harm was still right there after that period of its suspension. That we can fully rid ourselves of the fear of death or injury seems improbable for the body has a mind of its own and it never changes its mind. But if we can accept our death as an already-accomplished fact of a particular moment, we can be carried beyond our bodily fear of death, wherein lies a different worldview.

Decades after my injury-defying experiences I found the work of neuroscientist Paul MacLean on the "triune nature" of our brain, the subject of the first chapter of this book. MacLean's half century of research at the National Institutes of Health had revealed that we have within our heads three radically different brains and behaviors, including our basic body-brain and its compulsive survival strategies. Through MacLean's work I saw how fear of any kind throws us into an ancient survival mentality that, when fully active, shuts down our higher modes of evolutionary awareness. But it is these higher realms of our neural system that hold the open-ended possibility through which we can modify and modulate the reality structure of a particular moment. When Carlos Castaneda brought out his remarkable books, I saw that he clearly knew that our fear of death blocks us from using our full potential and the full spectrum of our humanity. Whether or not one accepts Castaneda's literary vehicles for presenting this fact is beside the point. Of significance is that he certainly knew about and must have experienced this phenomenon far more fully than most of us.

During this period of my twenty-third year, I took classes all day at the university and worked an eight-hour graveyard shift, six nights a week, running a bank check–proofing machine. I was doing poorly at both and was walking in my sleep until I discovered, from general desperation, that I could turn over the actual operation of the infernal IBM machine I ran all night to the now-familiar phenomenon of unconflicted behavior and it would run the machine for me. A check-proofing machine was a high-speed device on which I made frequent and costly errors, but through unconflicted behavior I could relinquish my post as its operator and sleep through my shift while the "force" of this phenomenon infallibly operated from the strength of my implicit trust. And sleep I did—quite genuinely, dreams and all—yet with eyes open and body busy as unconflicted behavior handled everything, even through the coffee breaks (which I didn't need).

The place where I worked was a bank clearinghouse and there were thousands of bank checks to process each night. Each operator was supposed to "close out" at every bundle of sixty or a hundred checks in order to "balance" or make sure no errors occurred either in our work or in that of the operator of the branch bank providing us with that bundle of items to process. An error of one cent would halt an operator's production until the inaccurate entry was found—even if it took all night and the following day. Because mistakes occurred often, an error checker moved up and down the row of machines to help trace them, though mistakes still slowed production. But suddenly I, a total novice to this work and a newcomer to the job, was running several thousand more items a night than anyone else, with no errors at all and perfect balances at the end of the shift!

Immediately I was seen as the boy wonder. What no one knew was that I never closed out at each individual packet of items as required. In fact, I didn't close out and check my balance until the end of the night when the shift was over, as to do so would awaken me, break the flow of things, and result in errors. Instead I continued operating straight through, and for something like three months I ran more items than anyone else without making any errors at all. This was nearly unbelievable to everyone, including my supervisor—the "force" did such superior work that I was given a raise.

Sleep, however, was my true bonus—and my carefully guarded secret until one morning when my supervisor discovered by chance that I had closed out and balanced only at the end of my shift. From his response you might have thought I had violated his mother! But an error among those fourteen or fifteen thousand items could have taken all day to trace out and though no errors had occurred, I was threatened with immediate dismissal if I did not close out regularly as required. My explanations were limp and unconvincing to say the least, and with anxious eyes following my work from that point on, I had no choice but to comply. The result was that I made errors, ran far fewer items, and slept through my classes all day.

Closing out was part of the greater issue of unconditional trust in my unconflicted behavior to run the machine—closing out would have been, in effect, to doubt, and such doubt would cause me instantly to revert to my usual conflicted state. Unconflicted behavior opens us to a freedom from doubt, but does so only when we are free of doubt of any sort to begin with—a true catch-22: Because unconflicted behavior occurs only when we are free of doubt, opening to and unconditionally accepting the state are simultaneous, not linear, events and so are not subject to any form of logic. That is, the sudden, intuitive hint of the actuality of unconflicted behavior was not like a question asking me if I was willing to allow the state or to go along with it. Rather, the opening of this state coincided with my instant acceptance of it without qualification.

Carlos Castaneda's metaphor of a "cubic centimeter of chance" suggests a rather wide margin for the nanosecond speed with which this opportunity opens and closes, almost like a single pulse we must fall through at the instant of its opening. This is why that greatest model of unconflicted behavior, Jesus, urged us always to be aware and awake—we never knew at what instant It, or He, or Whatever might come.

Another in this series of bizarre unconflicted events occurred at the Palos Verdes cliffs some miles outside Los Angeles, where I attended university. These cliffs were extremely high and virtually sheer, rising straight up from the ocean in a fashion similar to the far more stable cliffs north of San Diego, where much hang gliding takes place today. In addition to being so high and sheer, the Palos Verdes cliffs were "rotten," meaning that they were a loose conglomerate of shale, sand, and rock that made them extremely unstable—in fact, huge cave-ins occurred frequently, with large chunks of land falling into the ocean. Most of the area within fifty feet or so of the cliff edge was roped off with warning signs not to go beyond.

Eventually that whole section of peninsula slowly slid into the ocean and many enormous, elaborate homes were lost. At the time I was there, back in 1950, though, Palos Verdes was undeveloped, largely open, and a favorite picnic and hiking place. Once, friends and I chose to picnic right on the cliff's edge, ignoring the warning signs, as the young and foolish are apt to do. A friend and I hiked down to the ocean far below, using a long, winding trail some distance away and picking our way along the boulder-strewn beach to a spot that we deemed, correctly, to be just below our picnic spot above. My friend, knowing of my extreme vertigo (I had refused to go near the edge at the top of the cliff), jokingly challenged me to climb the cliff with him, even though it was almost vertical and obviously quite rotten, with nothing stable to grab hold of. Though I was terrified, in order not to be thought of as "chicken," I went along. We got no more than ten feet up when the whole section began to simply crumble and down we dumped, white and shaken and covered with sand and shale.

On looking at my friend's pale face, I sensed the familiar knowing inside of myself, that instant of being sure of what could be done if I threw myself away. "I'm going up," I said without fanfare, and started to ascend the cliff again, my friend shouting that I was crazy, that he wasn't responsible, that he wouldn't carry my body out of there, and so on. I simply started and kept going, my certainty absolute—I knew I could not fall or be hurt. Every handhold, every toehold collapsed under my weight and I could see nothing up ahead of me for the dust and debris that was falling from beneath my hands. I knew, however, that as long as I didn't stop, even for a second, to search for a handhold or foothold, all was well and I would continue to go up. I knew that any hesitancy or fleeting doubt would be the end of me. And this knowing gave me a most extraordinary sense of freedom and delight.

I went up swiftly amid a peculiar whistling that sounded around me, perhaps from the enormous gulps of dusty air my exertion demanded. I felt I was enormously powerful and enveloped by whistles, layers of sound that sustained me. At one point I glanced down through the dirt and dust and spotted my friend on the beach, a tiny antlike figure immeasurably far below. At that sight my exultancy grew to wild dimensions and I moved even faster. Shortly afterward, my feet and legs were suddenly no longer scratching and clawing into the cliff face as they had been—only my hands were now in contact while my body swung back and forth beneath my outstretched arms. I was not moving vertically; in fact, the cliff face was arching over my head toward the ocean behind me.

I had come to a large overhang formed by the roots of the scrub trees and growth covering the area. It was on this overhang that we had unwittingly made our picnic. As my body swung free I looked down and, seeing no cliff at all, just space, I experienced a most exuberant joy that spurred on my clawing and swimming up and out through the debris. What my hands found to grasp is a mystery to this day but suddenly I grabbed what I knew to be grass and then was up and over the edge. There before me were the others in our group, astonished, to say the least, over this apparition suddenly coming up from beneath them.

The bubbling of exuberance within me was now so intense that I was completely incoherent. I began to shout—a peculiar, screaming, animal-like cry of triumphant laughter that roared from my body without any volition or control. I was told later that I pounded the ground, pounded my chest, and made my animal noise for quite some time before growing quiet. By then my friend, seriously upset over the event, had reached our spot from the long roundabout trail.

The upshot was we all went back to the site the next weekend to settle the argument over belief or disbelief of my feat by checking out the seemingly impossible route I had taken. Some doubted their memory and the whole event when we viewed again that treacherous overhang from the vantage point of a neighboring cliff. My friend who had traveled the easier trail was subdued and silent, for indeed he had watched as I traversed the near-sheer cliff face in a veritable landslide of rocks and sand, and then, as I scrambled over some twenty feet of that reverse incline, going out toward the ocean as well as up. For my part, the sensation of my body swinging below my hands had been quite genuine—but the logic of the event just didn't add up.

In retrospect I realized that my wild, near hysterical elation was somehow connected to my acceptance of death in those moments, of taking death into myself, so to speak, so that I could in some manner go beyond it.

Following this incident, my next discovery was that an unconflicted person has dominion over a conflicted or divided person. Such dominion highlights the difference between the two types of behavior. As an unconflicted person, I was immune to danger or disaster during any unfolding event as long as I remembered to let the force of this behavior take over and avoided the knee-jerk reflex of fear and doubt. Miraculous or impossible events could unfold once I abandoned all hope and turned over matters to this peculiar force of will.

Again let me emphasize that this was never a negotiable decision. An instant's hesitation on my part erased all possibility—either I fell into the unconflicted state in the instant of its opening or I lost the chance. Further, the opening flashed to my awareness only in the actual context of an event, never beforehand. The perception of this opening and the decision to fall into it had to be simultaneous.

Interestingly, I found that I could initiate this state by arbitrarily placing myself in harm's way and maintaining my confidence that the opening would present itself at some critical moment when I needed it, as it had with the cigarette display for my dormitory friends. It seemed to be my confidence or freedom from doubt that brought about the revelation of that force, after which ordinary cause did not have to produce the expected effect.

Back in the early 1980s mathematician Ralph Straugh, author of *The Reality Illusion* (New York: Station Hill Press, 1983), having completed all the levels of aikido and four years of work with Moishe Feldenkrais in Israel, told me that no person can attack another without a deep, nonordinary agreement between aggressor and victim. After he mentioned this, I recalled Meister Eckhart saying: "Listen, when this birth takes place within you, no creature can hinder you." The birth Eckhart referred to was, in his words, the "birth of God in the soul," but there are undeniable similarities between Straugh's and Eckhart's point of view. There are many names for and facets of the shifts our spirit can bring about. Unconflicted behavior isn't a religious, theoretical, philosophical, or semantic issue, nor a matter of logic. Instead it is the alogical crack in the egg of reality, the way of faith, the way by which creator and created give rise to each other. Faith and belief are poles apart. Belief is intellectual and from the head. Faith comes, I can only surmise, from the heart, or perhaps from *kath* or *chi*—that center of will in our being.

The automatic dominion a person in unconflicted behavior assumes over a conflicted person brought matters to a head for me. I found that by shifting into unconflicted behavior I could sell anything to anybody. I dropped my all-night battle with the bank's IBM machinery and became a salesman purveying, of all things, sterling silver. Selling to poor, innocent working girls and struggling housewives, I made more in my first two or three weeks than I would have made in a year at my all-night IBM balancing act.

These extraordinary money coups began to bring a strong resurgence of that exultant exuberance I had felt during my cliff climb, and like a smitten gambler I began to play with the power, testing to see under what wild extremes the effect would work—and finding no discernible limits

Though this was over fifty years ago, I recall the final event of this long episode as clearly as though it happened yesterday. It was past midnight when I had run out of appointments with prospective customers and was heading home. I noticed that the neighborhood I was passing through was the same as the address of a prospect just given me by my last customer. I thought, why not stop and make one more sale? Who needs an appointment? So what that it was past midnight—give it a go! The thought of such a risky departure from my usual method sent my adrenaline and expectation sky-high.

I found the modest little house shut down for the night, not a light on anywhere, but I pounded on the door until a woman in her late middle-age opened it a bit to demand to know who was banging at such an ungodly hour. I asked for the name of the party given me and was told it was her daughter, long since asleep, after which the door slammed shut. I was filled with excitement and resumed my pounding. One thing led to another—why not to the police I will never know—and finally my magnificent display of silver was on their dining room table under full light while sleepy daughter in hair curlers, distraught mother in her bathrobe, and hulking but bewildered father stood by as I went on pell-mell with my sales pitch. The irate mother kept screaming at her husband to throw me out: "Throw this little mouse out! Throw him out of the house! What's the matter with you?"

At each new outburst from the woman sheer exhilaration and excitement welled up in me even more strongly and I began to laugh until tears streamed down my cheeks. I knew that they couldn't lay a hand on me, that I had them. The more I laughed, the more angry the mother became and the more bewildered the other two looked; the more surely they all lost control, the more vulnerable they were. The sale was a foregone conclusion.

Now the odd thing was that when I left in due time with the down-payment on a large order in my pocket, both mother and father walked me to the door, arms around me, and begged me to come back and visit them. This peculiar twist had occurred before under less extreme circumstances, but this one was my undoing. In retrospect I saw that the average person in his or her conflicted state of uncertainty, doubt, and fear—which was my ordinary state as well—was not only powerless in the face of unconflicted behavior but also seriously attracted to this state. Beneath these individuals' reactions of anger and frustration, a longing within them had been touched. This realization brought a whole new aspect to this already new perspective. It became clear that for me this unconflicted behavior was a common version of those famed temptations in the wilderness (if I may place myself and my petty affair in such great company)—and I knew from my gloating exuberance over the power I wielded that I had nowhere near the personal character or wisdom to handle such a force.

So, not from noble virtue or lofty principle but from fear and trembling, knowing that I was hopelessly out of my depth, I quit selling and resisted any temptation to monkey around any further with the "ontological constructs" of my world. Eventually I took another job, held on somehow at the university, and played it straight. Eventually I also lost my intimate contact with this opening and its alternate reality, though it has always been in the background of my mind, making me question our common consensus of what is possible and what is impossible. It was this that led me several years later to begin writing my first book. All I have written since, including the following chapters, has been but a sequel prompted by the enigma of unconflicted behavior, for in this phenomenon lies the key to who we are and what we can do to find our transcendence and escape the current violence we bring to ourselves and earth.

I now turn toward understanding both how this key works and that which lies beyond the gate it opens.

ONE

▼

EVOLUTION AND OUR FOURFOLD BRAIN

*When the higher incorporates the lower into its service, the nature of
the lower is transformed into that of the higher.*

—MEISTER ECKHART

The size of the Sphinx of Giza in Egypt is astonishing. Almost as long
as a football field and as tall as a six-story building, it is the world's largest
monument carved from a single stone. According to the archaeological studies of Schwaller de Lubicz reported by John Anthony West in his book
Serpent in the Sky (New York: Harper and Row, 1979), it was carved many
millennia before the first Egyptian civilization. Discovered first by early
Egyptians in about 6000 B.C., the monolith was forgotten and rediscovered
time and again over the following millennia—unless continually cleared,
the great basin in which the giant sculpture and its complex rest soon fills
with wind-blown sand, ultimately burying the Sphinx and leaving only the
topmost part of its head exposed. In about 1400 B.C. the Egyptian king
Tuthmosis IV dreamed of its presence there under the ocean of sand and
had it uncovered and brought to light again, thus beginning anew the cycle
of burials and resurrections.

In the eighteenth century members of an early scientific society in
England made superb drawings of the entire Sphinx complex. Later, toward the end of the Napoleonic Wars, a scruffy bunch of Malumek strag-
glers came along and used the head of the vast monolith as a target for their
little brass cannon. This was followed in the early 1800s by a group of
scientists and artists commissioned by Napoleon to study and make detailed

drawings of the remains. Other than the artillery damage inflicted, the French drawings and diagrams show the same figure and features described by the English.

Until damaged, the Sphinx was a composite creation that symbolized the three major periods of evolution on this earth and, for close to twenty thousand years, embodied in stone our own evolution. The main body of the giant creation is that of a lion, king of beasts in myth and legend. High on the beast's chest, nestled between outstretched paws, are human breasts, while rising from the body is a masculine human head, his countenance gazing slightly upward as though at a distant horizon. The effect of the serene pose is ethereal. Added to these three was one more element, though we have only drawings of the sculpture in its intact state: Until modern man and his brass cannons arrived, there arose from the top of the skull, as the crown of all this magnificence, a giant hooded serpent, curving with artistic grace until its head rested just above and between the great, gazing eyes of the massive human head beneath.

The Malumeks blew away part of the brow, cheek, and one eye, and most of the nose, including the bridge, and left almost no trace of the great serpent. In so doing they gave us a graphic update of our human story: magnificence reduced to a broken figure and a sorry tale. For now, we must bear in mind what the original Sphinx represented: the body of a lion, human female breasts, and male head adorned with a great reptile rising out of it—a unity of reptilian, old mammalian, and new mammalian, the human experience in stone.[1]

---

1. In the Hindu tradition the hooded cobra represents the Kundalini Shakti, or universal creative energy from which life springs, and is a mediator between Shiva, the supreme god, and mortal man. Shakti's movement up through the great cycles of evolution culminates in human life. In this tradition the crown of our head is the seventh chakra. *Chakra* comes from the Sanskrit word for "wheel"; these areas of the body are considered whirling wheels of energy and may have some relation to neural centers. When the serpent power breaks through the bounds of skull, and thus of mind or thought, it brings enlightenment, the highest state of life. In the version presented by the Sphinx, the great Shakti then curls over and rests her hooded head right in the center of the forehead at the ridge of the brow.

THE STRUCTURE AND FUNCTION
OF OUR THREEFOLD BRAIN

▼

For decades neuroscientist Paul MacLean was head of the Department of Brain Evolution and Behavior at the National Institutes of Health, one of the great research centers of our time. His extraordinary work spanned six decades—he was still producing brilliant papers in 1997—and was in part based on his discernment of a striking similarity between the three neural systems in our head and the brain structures of the three major animal groups of evolution: reptilian, old mammalian, and new mammalian. For more than a half century he and his staff traced these parallels and showed how each of our neural systems carries within it the blueprint of potential intelligences, abilities, and capacities developed during each of these evolutionary epochs.

Nature never abandons a system that works but instead builds new, enlarged, and more efficient systems upon the old. She seems to have created each new evolutionary brain to correct problems in an older system or to expand its possibilities. To the three inherited neural blueprints we add the content of life's ever-changing environments from which arises our extraordinary adaptability. The striking differences among the three neural structures of our brain make this heritage both a blessing and curse, however. When integrated, these three systems offer us an open-ended potential, an ability to rise and go beyond all constraint or limitation. But when that integration fails, our mind is a house divided against itself, our behavior a paradoxical civil war—and we become our own worst enemy.

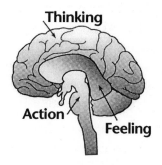

Figure 1. Courtesy of Touch the Future Foundation.

In the remainder of this chapter we will look at the structure of this evolutionary brain as identified and examined by MacLean, focusing on the function of these parts as they work together—or separately.

Neuroscientists originally divided the human brain into a simple hindbrain and forebrain—still a useful description. Our hindbrain is the reptilian brain (called the R-system by MacLean) consisting of our sensory-motor system—the spinal cord, the body's vast network of nerve endings and their neural connections, and the primary neural systems in the heart. The forebrain is made up of the old mammalian brain and the new mammalian brain (the neocortex).

Before moving forward to discuss how our brain came to be as it is now and how it functions as a unit, following is a short summation of its first three parts. (Our complete, fourfold brain includes the prefrontal cortex, whose story of development requires a chapter of its own—chapter 2.)

RITUAL, HABIT, AND THE ROOTS OF LYING: THE REPTILIAN BRAIN

Our R-system functions in a habitual, patterned way and is unable to alter either inherited or learned patterns of behavior. This ancient brain, however, can take over the physical parts of a learned skill such as typing, bike riding, driving a car, or the sensory-motor aspects of playing a piano, thereby freeing our new brain to stand outside of the immediate motor action and observe and discover ways to improve or perfect the performance.

This oldest of our four neural structures is skilled in deceptive procedures that were developed eons ago to elude predators. These allow us to "change our color" like a chameleon according to social environment and also enable us to be not just two-faced but instead multifaceted, particularly if we feel threatened in any way. This deceptive skill can be used on behalf of our high neocortex to develop strategies for succeeding in the worlds of commerce and politics, for instance, after which our high neocortex can skillfully rationalize and make morally respectable, at least to ourselves, our often quite immoral actions. Through the alliance of our neocortex with this deceptive low brain we learn to lie, gloating gleefully when successfully deceiving, lamenting and self-pitying when so deceived.

Besides generating these survival strategies, our reptilian R-system handles, beneath our awareness, numerous decisions about our physical wellbeing, many in tandem with other parts of the brain. In emergencies this reflexive system can alert our third brain, the neocortex, to an event that needs quick attention and possible mobilization of all systems for the body's defense. Through an *interpreter mode* in our verbal-intellectual neocortex, we can make swift decisions and mobilize our intellect to the R-system's defense network, unimpeded by emotions or any other consideration that might interfere with a quick reaction to the emergency.

The quick, reflexive reaction built into the R-system can be handy in a dogfight—but there are situations that require a more integrated, whole-brain approach. The same "emergency" reports can be passed up to the neocortex through our emotional-cognitive system (or old mammalian brain), which mediates between the R-system and the neocortex. Through this broader connection the interpreter mode in our neocortex can stand back and moderate, monitor, or even redirect sensory reports. This makes for a measured and more creative approach to what might otherwise be a violent reaction were our reptilian brain functioning on its own.

THE OLD MAMMALIAN, LIMBIC, OR EMOTIONAL-COGNITIVE BRAIN

We call our second neural structure the old mammalian brain, and it is indeed quite similar to that found in all other mammals, as are the behaviors and abilities apparently encoded within it, such as our inherent intelligence for nurturing our young. Because this structure surrounds the basic R- system like a limb, we call it the limbic system. It is also termed the emotional-cognitive brain, for here nature adds to the reptile's limited senses our extraordinary senses of smell and hearing, which lift the whole sensory system to a new order of functioning and open an entirely new world.

Additionally, here in this nurturing emotional brain are the foundations for all forms of relationship, including our general cognition of the world as somehow "other," as something to which we must relate. A reptile's relationships are simple: When its primitive vision spots a moving clump of contrasting light and dark (the only visual discernment it can make), the reptile asks, "Is it something to eat, mate with, or be eaten by?" Thus the repertoire of its subsequent actions can be classified in two ways: Go for it or get away from it. The mammalian system is infinitely more complex than this, and infinitely more discriminating. The collective term for those tools by which we qualitatively evaluate all our relationships—particularly our relationships with each other—is *emotion*.

Although the reptilian and old mammalian brains support, mutually influence, and interact with each other, nature certainly made a quantum leap beyond the reptilian in developing the old mammalian, with no discernible transition. In humans, the R-system gives us awareness of an outer, sensory world while our emotional brain gives us awareness of an interior, subjective world and our feelings concerning that outer world and our relation to it. The marked change of behavior between the infant pushing itself around on its belly in reptilian mode and the standing toddler is brought about by a developmental shift of focus from our sensory-motor R-system to the emotional-cognitive, or limbic, brain. From its new perspective, the toddler can stand "outside" his sensory-motor brain and begin to employ the far more evolved and sophisticated capacity to relate to his or her world as an object, rather than simply act reflexively to sensation.

THE NEOCORTEX, NEW MAMMALIAN, OR VERBAL-INTELLECTUAL BRAIN

Our third brain (neocortex) introduces language and thinking, the ability to stand outside all other activities of the brain and observe these activities objectively and consider all factors of a situation rather than react to them from instinct alone. This high brain occupies five times more skull space than the reptilian brain and the old mammalian brain combined and consists of some hundred billion neurons. Each neuron is capable of interacting with upward of a hundred thousand other neurons to form fields of coordinated neural action. These neural fields translate particular frequencies to our awareness and our awareness to other fields, all of which field effects are constantly shifting and changing, updating their various intelligences and reports. There are no limits to what our third brain can translate, from input from the world out there to imagination and thought within. With the development of this third neural structure, nature opened up an infinitely wide window of awareness.

The first brain registers present tense only; the second computes both present and past. With the addition of the third brain, we possess awareness of the past, present, and future. But evolution here runs into a snag even as an entire new universe is opened. The future introduces the "What if" syndrome—What if the sun goes out—Who's got the flashlight? Herein originates the useless anxiety and concern that can be brought about by this forward-thinking brain—a feeling state that can result in this new mammalian brain being pulled into the service of our lower survival ones.

This ability to conjure the "What if" scenario is not the only problematic function of the third brain. Its impulse toward novelty is one of our most intriguing drives. A cat's curiosity is nothing compared to ours—through this higher brain we are motivated toward continual expansion of our awareness and experience. Our drive toward novelty is a tool of evolution and transcendence: Evolution may have exhausted its possibilities for (or interest in) novel flora and fauna on this good earth, but through our intellectual-creative brain it introduces creative imagination, which is the foundation of all organized thought and creative intelligence. Once stirred into action, thoughts boil forth endlessly from this wild frontier of imagination. Unknowns of every conceivable form and universes of weird, improbable notions and fantasies spill teeming over its boundaries. The medieval Sufi spoke of the imaginal worlds and considered imagination the highest human capacity, the way in which we are most Godlike. This observation was similarly made by Jacob Boehme, William Blake, Goethe, Rudolf Steiner, and other great beings of our history. God imagines us in order that we might imagine him—image to image, mirror to mirror, the creator-created dynamic.

Whether we follow novelty's call to adventure or close ourselves in a defensive posture, refusing to engage, depends largely upon the experiences we have in the first three years of life. These years mark the time when our emotional system develops the foundation for our higher intellect yet to come. And that's why Jesus, our great model, said that if we "cause one of these little ones to stumble, it were better a millstone be tied around our neck and we be dumped in the sea." It is thus we are a drowning species.

THE EVOLUTION OF OUR BRAIN:
INCORPORATING OLD INTO NEW

▼

Our three brains develop in utero as a nested hierarchy in the order of their appearance in evolutionary history: The reptilian brain begins its functions in the first trimester of gestation, the old mammalian in the second, and the neocortex, or human brain, in the third. Nature's newest addition, our prefrontal cortex (prefrontal lobes), makes its major debut after birth. (See chapter 2.)

Although the brain's older neural modules or parts are similar to those in other animals, the overall context in which a module is situated determines its capacity and function. The spark plug in my chain saw serves essentially the same function as the spark plug in my neighbor's Mercedes-Benz, though the overall difference in performance is striking. Neural environment, the aggregate of different brain parts, is a major factor in determining how a particular part will function.

Nature's pattern of development is itself threefold. First, each new neural structure is built on the foundation of neural structures that have come before it. Second, as each new brain develops, it incorporates into its own functions the more primitive foundation upon which it is built and changes the nature of that foundation into one that is compatible with the new system. And third, the newly integrated system serves, in turn, as a foundation for higher evolutionary developments, which is transcendence in action.

Biologist Bruce Lipton shows how the first cell created by nature was, in effect, a brain unto itself and a template that underlies all subsequent development. The neuron as a specialized cell first appeared singly, organizing a small group of those lower-order cells into a new order of coherent action. This "smart" cell, the neuron, became the boss, or manager, of simpler cells preceding it, sending them its orders over a slender neural thread. Out in the fishpond, for instance, there dwell extremely tiny wrigglers called hydrochondria, which are food for the littlest fishes. Each of these minuscule creatures consists of two rows of about a dozen cells each, every few cells being a neuron, which connect to one another by a slender neural thread running between the two rows. Through this communication link the smart cells coordinate the whole entourage, whether wriggling about in pursuit of something to eat or avoiding being eaten.

Beyond the hydrochondria's simple neural construction the organization grows more complex. Eons ago, neurons gathered in groups called *ganglia*, connecting with one another through more elaborate threads called *dendrites* and *axons*. This organization presented ever-greater possibilities for group perception of an environment, along with more elaborate actions for survival. Ganglia led, after a few more eons, to such enormously complex ganglia groupings as the brains of reptilian amphibians. The reptilian brain became, in turn, the foundation for the earliest mammalian brain, which became the foundation for the even more advanced new mammalian brain, leading to our present neural structure.

Incorporation of a previous system into a new one changes the earlier function to one that is compatible with and supportive of the new. When that sophisticated cell the neuron appeared and incorporated all cellular functions into its own operations, it changed the behavior of all other cells accordingly. As evolution led the reptilian brain to join with the emerging old mammalian, essentially the same old R-system existed—but when functioning in synchrony with the mammalian brain, the synchrony of the reptilian brain operates quite differently from when it was the only game in town. The R-system still governs the sensory-motor and survival systems but is also subordinate to that even smarter, more complex mammalian system in a correspondingly larger environment.

When the neocortex or new mammalian brain came along, each of the other two became subordinate parts of this even larger organization, though still retaining their respective jurisdictions and responsibilities. In turn, this triune brain paved the way for—and was designed to serve—a fourth brain, our prefrontal cortex, so called because it is attached to the front part of the neocortex (the area just behind our brow). This fourth and largest system should reign over the three existing structures, but breakdowns in communication or even mutiny in the ranks—always a possibility in such hierarchies—does occur. This, however, gets ahead of our story.

Intriguing to our human aspirations is that the essential nature of any older system, when integrated into a newer one, retains its integrity while playing its new, expanded role. The old mammalian brain, when it came along, changed the nature of the reptilian brain onto which it had been grafted, but that R-system still functioned for survival, albeit in a more intelligent, flexible, and adaptive way.

This process, however, is not a one-way street. All systems are dynamics that move in two directions—between the old and new—so that some of the essence of the higher is absorbed by the lower even as the lower is itself incorporated into the higher. Each brain, preceding and emerging, modifies the other to some extent. Thus, the resulting transcendence, or movement beyond limitations, is not reached at the expense of the unique achievements of each system incorporated. If this were not the case, both nature's economy and the transcendent aspect of evolution would be defeated. In our case, for instance, in order to transcend our present state we must be incorporated into a higher order of operation. But individuality itself is what is lifted up into that new order, for an individual self was (or is trying to be) the unique achievement of our particular stage of evolution.

THE INDEPENDENCE AND INTERDEPENDENCE OF OUR THREE BRAINS

▼

The reptilian brain was hundreds of millions of years in the making and perfecting. The foundation for all subsequent brain evolution to this day, this neural structure gives us and all other mammals our sensory-motor system, with its association with our physical body and a rich heritage of survival and maintenance instincts. All neural developments since have depended on this ancient foundation. We couldn't have evolved or survived without our reptilian ancestor setting up shop in our basement—nor could we survive as humans were that original reptilian temper not modulated a bit by the higher structures built upon it.

This duality—the independence and interdependence of each of our neural structures—can cause trouble should their integration or entrainment fail. Failure to integrate can lead this cantankerous trio—reptilian id, mammalian ego, and neocortical superego—to erupt in near constant scraps over who gets to play king of the mountain. Indeed, confusion over which of the three gets to integrate the other two into its service is the source of all the cheap theatrics cluttering life's stage in what should be a great ongoing drama.

LEARNING AND MEMORY

Our emotional brain is the seat of all relationship and is involved in memory—recalling what we know. We learn by relating something unknown to something we know. Even the abstract capacity for associative thinking, whether scientific, mathematical, or philosophical, though dependent on our third brain, has its foundation in the feeling state of the old mammalian brain.[2]

---

2. Antonio Damasio explored this in his book *Descartes' Error* (New York: G. Putnam and Sons, 1994).

Between our emotional-cognitive (old mammalian) and sensory-motor (reptilian) brains lie two critical modules, the amygdala and hippocampus. The *amygdala* is involved in recording our earliest emotional and survival experiences and learning in the first three years of life, an activity that functions beneath our awareness thereafter and largely shapes the way we respond to events. The *hippocampus* experiences its period of growth after the third year and is involved with general memory from moment to moment and any transfers to long-term memory. Its operation too centers on survival strategies and relationships.

Our second, emotional brain relates directly with the temporal lobes and right hemisphere of our third brain, or neocortex. Dreaming, intuition, creativity, and related phenomena take place as a result of this interaction. These two higher brains, emotional-cognitive and verbal-intellectual, can join forces to alter the basic R-system functions that give us our experiences of body and world. For instance, a capacity called *concrete operation* begins to develop at about age seven and through it we can "operate" on physical phenomena, meaning that our perceptions from the R-system can be changed by an abstract idea from these two higher brains. We can use an abstract idea to intervene in the natural processes of the lowest sensory-motor brain. Through such concrete operations of mind in which the higher brains incorporate the functions of the lower brain into their workings we can change our experience of the world. We can, for instance, imagine new ways of keeping warm in winter or walk without harm through pits of fire that would melt aluminum on contact.

Neither our sensory-motor nor our emotional-cognitive system functions in us as it does in animals, though the neural structures and behaviors associated with each are similar. The presence of the neocortex transforms their nature—if this higher neural structure is developed. Interestingly, failure to develop this highest brain lies most often in a failure to develop its foundations, the old mammalian and the reptilian. Such early failure leads to an unending cycle of breakdown in the dynamics between the neural structures. We can modulate the lower, instinctual reactions of our survival system through our neocortex. But our high brain neocortex can be developed only on the firm foundations of a well-developed survival brain. If we fail to develop the reptilian brain sufficiently, the neocortex can't integrate the R-system into its service and modulate its behaviors. When the ancient reptilian brain dictates behaviors without the modulating or tempering of the neocortex, trouble brews for that person, his or her society, and the larger body of the living earth.

THE PROXY NEOCORTEX AND THE MODEL MIND

Because an infant's emotional-cognitive brain can't develop until his sensory-motor system is fully functional (after the first year of life), the mother is programmed by nature to act as the infant's old mammalian brain, establishing appropriate relationships, nurturing, stimulating, protecting, during the first year when the R-system's primary development unfolds. Once the infant's R-system functions with some independence, the mother or primary caregiver then acts as the model for the development of the toddler's forebrain (old mammalian brain and neocortex), beginning with the emotional-cognitive system. Because the child's verbal-intellectual brain can't develop until a functional emotional-cognitive system is in place, during this period the caregiver is also the child's proxy neocortex as well as the model for the child's development of his own verbal-intellectual brain. This important role of the caregiver demonstrates nature's *model imperative,* the procedure by which all development takes place. (For further explanation of the model imperative, see chapters 2 and 6.)

If nurturing is complete and development is successful, the child can monitor and control his own sensory-motor and survival systems with increasing independence as he grows. This is called *affect-regulation* in classical psychology, or *emotional intelligence* in popular parlance, a subject that will be explored in part 2 of this book.

Once it is functional, our emotional or limbic system (old mammalian brain) can organize its two neighbors, the lower R-system and the higher neocortex, into focused attention (concentration on what is to be learned). A positive emotional state *entrains,* or unites, our systems for thought, feeling, and action; shifts our concentration and energy toward support of our intellectual and creative forebrain (old mammalian and neocortex); and allows us to both learn and remember easily. In very young children, the primary caregiver's emotional state determines the child's state, and therefore the child's development in general.

On the other hand—and this is true throughout life—any kind of negative response, any form of fear or anger shifts our attention and energy from our verbal-intellectual brain to our oldest survival brain. In such instances, we don't have full access to evolution's higher intelligence and react on a more primitive level. When we are insecure, anxious, undecided, and tense, the focus of attention can become divided among the three brains, each with its own agenda, so that we are thinking one thing, feeling something else, and acting from impulses completely different from either of these. In this all-too-common confusion, children's learning and development are impaired and adults' decision making and thinking become faulty.

This brings us to a major dictum of nature: A negative experience of any sort, whether an event in our environment or simply a thought in our head, brings an automatic shift of attention and energy from our forebrain to our hindbrain—that is, away from our higher verbal-intellectual brain toward the lower R-system and its defenses. This shift shortchanges our intellect, cripples our learning and memory, and can lock our neocortex into service of our lowest brain.

Whether the signal received by the emotional-cognitive brain is from higher up the evolutionary stream (from the neocortex) or from lower down (from our sensory-motor system), our emotional response is the same. A negative signal from either direction brings a negative response throughout the emotional system, which is then reflected throughout the entire body and brain. Just as the emotional-cognitive brain dutifully responds to alert signals from the R-system telling us, in effect, that a saber-toothed tiger is coming, so our second brain responds to evaluations, criticisms, fears, and anxieties brewed up in our creative imagination—and in our creative brain we can imagine a thousand different ways that saber tooth might come! The cause, accuracy, or validity of our negative imaginings makes no difference: Any negative thought or event brings a shift of energy and attention from our forebrain to our hindbrain and does so completely beneath our awareness.[3]

STATE-SPECIFIC LEARNING

Years ago, research people discovered that the emotional state we are in when learning takes place becomes an integral part of that learning. We call this *state-specific learning*. Candace Pert's discovery of the "molecules of

---

3. I have dwelt repetitively on this negative aspect here simply because herein lies our general downfall.

emotion" shows how all hormonal function, including that of the immune system and even allergic responses, occurs as a sophisticated memory system handled primarily by our emotional brain. Any negative emotional reaction or state triggers a corresponding flood of specific hormones. Because learning and memory are emotional-cognitive functions, the neural pattern, imprint, or "structure of knowledge" (to use Jean Piaget's term) of a specific learning event includes in its content the memory patterns of those emotional hormones prominent in the body at the time of that learning. Thus the emotion experienced while learning something becomes part of the learned pattern. When we exercise that learning, even years later, the same emotional hormones will fire on cue, for they are as much a part of the neural pattern as is the learned material—and our body, brain, and heart respond accordingly.

If we learn our arithmetic "to the tune of the hickory stick," the pain and fear of that stick are as much a part of our knowledge as are those numbers. The result is that we may find ourselves reluctant to recall what we actually know—on some deep level numbers are associated with fear or pain, which moves one part of us to protect another part from a repetition of that trauma. In such a situation, any mathematics can be traumatic and recall and learning in the subject impaired.

WHEN ALL THE BELLS OF HEAVEN RING: THE REWARD CENTER

At the other end of the scale, decades ago James Olds discovered a "reward center" in the emotional brain that accounts for the high level of pleasure we experience with certain relationships and actions. For instance, this reward factor links with the R-system's drive for species survival and our highest brain's novelty factor to produce human sexuality. The results bear scant resemblance to the simpler drives of our ancient kin; they absorb our attention to a remarkable degree and produce some astonishing and outlandish behaviors. This reward center may also play a role in ecstatic seizures, raptures, and mystical experiences.

In one of Olds's experiments, an electrode was implanted in a human subject's reward center. Upon the electrode's activation, the individual reported "all the bells of heaven were ringing." This shouldn't imply that our reward center contains the makings of ecstasy or rapture—the experience doesn't dwell there, waiting to be prompted to expression—but the center may provide the modus operandi through which we perceive such a state. Through this particular neural grouping cued to a certain frequency response, a highly specialized form of experience can be translated. The final experience can't be "found" anywhere; it exists only in the overall function of this center (the fluid frequencies translated by its neurons) and its corresponding interactions (our fluid, not localized, reception and reactions).

Recall Meister Eckhart's claim that there is no being except in a mode of being. If I tune my radio receiver to a certain frequency, I get a certain set of information. But those transistors in that radio do not contain within themselves those frequencies or that information. Likewise, there are no grandmother cells in the brain containing blocks of preset information or fixed content, only neural networks to interpret corresponding frequency networks. A single neuron might act as a lone target cell, which can activate a whole network of cells for a particular perception or memory. The target cell doesn't contain the perceptions of the field it targets, but it can trigger a specific pattern of neural fields that are involved collectively in some selective action or memory.

Olds found that rats with electrodes implanted in their reward centers would forgo food to the point of starvation and water to the point of death from dehydration, would ignore a female in heat, and would even cross an electrified grid in order to maintain contact with the electrodes that activated the center. (Rats are not ordinarily prone to celibacy and reportedly will avoid shock at all costs.)

Considering that most mammals and avians possess a reward center, I am reminded of William Blake's quatrain question:

> How do you know but every bird
> That wings the airy way,
> Is an immense world of delight,
> Closed to your senses five?

Our emotional (old mammalian) brain gives us our immune system and monitors our hormonal systems, our body's ability to heal itself, our body's biorhythms, our personal relationships, our preferences, and our general aesthetics. Here are those herd instincts found in most animals that should blossom into our social worlds of shared experience. All of this action rests on our capacity to remember, and each brain makes its own contribution. The cyngulate gyrus, largest module of the limbic system, connects to the third brain (neocortex) on many levels and has direct links to both the amygdala and the hippocampus, each of which is intricately involved with the function of memory. (The amygdala, you will recall, is the repository of long-term survival memories from the first three years of life; the hippocampus controls the memories of the later years.)

THE LOOSE CANNON: THE LEFT HEMISPHERE

While our emotional or old mammalian brain has rich connections to the right hemisphere of our third brain, direct connections between the emotional brain and the left hemisphere are sparse. Herein the plot thickens and grows hazardous. Having no efficient connections of its own, the left hemisphere's relation to the rest of the three-brain system is largely through the right hemisphere, which has much stronger connections to the brain as a unit.

A thick neural band called the corpus callosum bridges the brain's two hemispheres, and it is through this that the left hemisphere has access to the right and to the processes of the rest of the brain. This neural connection allows the left hemisphere to take material from the right, retreat from the unified scene, and operate on that material in isolation—dissecting it, analyzing it, and putting it back together in novel ways without regard to the checks and balances of the rest of the system and without concern over how this new creation will relate to the whole. (The left hemisphere, however, has a connection to the prefrontal cortex, as we shall see, and it, in turn, is intimately connected with every facet of the brain.) So through all this action, the left hemisphere maintains its dynamic interaction with the prefrontal cortex, from which creativity and novelty spring. Such isolated left-hemisphere maneuvers, though, are not available to the holistic right hemisphere, connected as it is to the emotional-cognitive brain to which, as a result of this connection, the right interacts.

Paul MacLean and his friend the late Arthur Koestler, who wrote on this anomaly, considered nature's failure to knit the left hemisphere firmly to the emotional brain a fatal error because the left hemisphere, with its maverick behavior and seeming disregard for the rules, might be responsible for creations such as bombs and catastrophes such as holocausts. But the recent research of Elkhonon Goldberg indicates that nature didn't err after all and, as we shall see, many assumptions concerning left/right hemisphere differences, so popular over the past twenty years, may be, at best, half-truths.

Until recently, the right hemisphere, although verbal, intellectual, and creative on its own, was largely thought to work holistically to maintain a balanced unity of all three of our neural systems. The left hemisphere, as a later evolutionary addition than the right (it even forms later in utero), was thought to be more powerful and able to override the right hemisphere, even though the right is the left's source of information for functioning. Because of these attributes, the right hemisphere was considered feminine and the left masculine, an assumption questioned by Goldberg's research.

He claims the real difference between left and right hemispheres is that the right hemisphere is adept at handling novel material while the left seems to be the repository of all fully developed structures of knowledge, handling all learning that is stabilized and firm. The right hemisphere, with its rich connections to the two lower brains, is involved in new learning, and when this learning becomes fairly established the left hemisphere takes over. With its command over all stable patterns of knowing, the left hemisphere, in its comparative isolation and in conjunction with the prefrontal cortex (our most recently developed fourth brain) and the information it has gathered from the right hemisphere, can move into logical, analytical, associative, and creative capacities without being overly influenced by the more primitive lower brains. That is, the left hemisphere can function outside the boundaried conditions of the two lower brains, thus avoiding the more knee-jerk strategies of the R-system. Without the lower brains' boundaries, we can move beyond our ancestry and thought can take wing.

EGO, INTERPRETER, AND INTELLECT

The left hemisphere's predilection for novelty and intellectual adventure without regard to well-being or balance is a key feature of ego-intellect and its *interpreter mode*. The interpreter mode was proposed by Michael Gazzaniga, who claims this is a module within the otherwise nonmodular neocortex and that its hallmark is the ability to stand outside of any information, look at that information objectively, and interpret it according to general conditioning or learning. The interpreter can lead to brilliant creative thinking, yet can be devoid of intelligence, intelligence being a generalized move for well-being that is generated by the heart and limbic systems and their connections with the right hemisphere and prefrontal lobes. (We will touch on this again later.)

The left hemisphere and its hypothetical interpreter have no direct connections to that central intelligence agency made up of the heart and emotional systems, as our daily news informs us. It is thus that there is no trace of intelligence in those creating a neutron bomb—but ah, what brilliant intellect is there!

A vast creative potential opens through the connection of the hypothetical interpreter mode and the well-known left hemisphere independence. Through this freewheeling connection we can play intellectual games without reference to any previous evolutionary system. Thought can break free from all its inherited animal constraints and move into the unknown, and with this move creates the very terrain it discovers—for it is at this point in the creator-created dynamic that we are unknowingly approaching creation itself. Here in our drive for novelty, our curiosity, and our compulsion to investigate and push beyond boundaries, we are evolution itself, expanding into new openings.

Though this may well be what creation had in mind, the moderating or modulating intelligence of well-being and ecological balance can get lost in this shuffle. Therefore, the balancing, integrated, right-hemisphere intelligence—which seems related to the feminine—must be firmly established as the foundation to our great venture into creative thought lest we self-destruct. Happily, this is precisely what nature provides for in her developmental stages first explored by Rudolf Steiner, then by Maria Montessori, Jean Piaget, and a steady stream of bright researchers whose findings are too often ignored.

BEASTLY KINGS AND FROG PRINCES

▼

By way of summary, a higher system can incorporate a lower one into its service and can transform the lower. Or the converse can happen—the higher can be incorporated into the lower and the transformation is reversed, inverted, with the lower transforming the higher. In our legends and fairy tales, the magnificent king makes an error or acts in a base manner and is turned into a beast, or the handsome prince is made a frog, and in both instances their high human cortexes are incorporated into their animal brains—and lord help us! Or rather, lady help us, for she alone, through her nurturing kiss and inherent wisdom, can bail out the male from the predicament his intellect constantly brings him to.

The creator-created dynamic of our nature centers on two pivotal areas of evolutionary development: the heart thumping away in our chest, one of nature's prime and primary achievements; and the prefrontal cortex, there behind our brow. The solution to our violence and our key to transcendence lies in translating nature's newest creation and most fragile experiment, our fourth brain, into our own experiential venture—and in the next two chapters we will explore the biology behind this journey.

In this chapter I have outlined the barest sketch of our brain-mind, the most complex organization known in the universe—but this spare sampling must suffice for our exploration to follow. To give us our worldly experience, this vastly complex brain, itself an intricate balancing of frequency fields, translates from even vaster fields that are infinite in extension and dimension. More and more, all experience is being interpreted by the new science as an infinitely interwoven complex of resonant frequencies that selectively modulate to form worlds, bodies, thoughts, emotions, rocketships, and sonnets. More and more researchers say in a mounting clamor of consensus: "It's all just frequencies!" Just frequencies! If they are right—and they seem to be—our mystery has just grown immeasurably deeper. Follow the dynamic of creator-created far enough, however, as I intend to do here, and a workable solution to this mystery will emerge, along with a key to transcendence and its dark opposite—violence.[4]

---

4. Admittedly a far simpler and more congenial cosmology-theology gave us Jehovah sitting on cloud nine surrounded by hosts of angels who watched as he threw miracles and creations about. Here was God as the biggest object in a clearly defined objective world. Those, in fact, were the good old days.

▼

TWO

EVOLUTION'S LATEST:
The Prefrontal Lobes

Intent precedes the ability to do.

—JEROME BRUNER

Immediately behind the ridge of our brow lies the prefrontal cortex (the prefrontal lobes), the largest and apparently most recent of brain additions.[1] Behind the prefrontals lies the rest of our neocortex. While our reptilian brain has modules or parts that are hundreds of millions of years old, indications are that only about 40,000 years have passed since our prefrontal lobes appeared in their present size and with their current significance.

The curving serpent that once crowned the great Sphinx of Giza rested its head at the very center of this prefrontal area. In Eastern spiritual systems this center of the forehead is variously called the third eye, guru chakra, ajna chakra, and sixth chakra, a chakra or "wheel" being a center of energy. Eastern and occult philosophies claim that the awakening of this area of the brain can open us to higher worlds—the imaginal worlds of the Sufi,

---

1. Classical evolutionists refute the theory of the recent appearance and expansion of the prefrontals, claiming that traces of them can be found as far back as the Big Bang. See Harry J. Jerison, "Evolution of Prefrontal Cortex" in *Development of the Prefrontal Cortex: Evolution, Neurobiology, and Behavior,* Norman A. Krasnegor, ed. (Baltimore: Paul H. Brookes, 1997). There are items overlooked in this reactionary view and I will stick with Paul MacLean and his evidence of recent prefrontal development, which seems the current consensus. Traces of them can be detected in any mammalian brain and they are found in the higher apes, but these are nothing like the size and scope of ours.

for instance—and to various subtle realms of possibility, all of which may indicate aspects of prefrontal function.[2]

Neuroscientists have a variety of viewpoints on this comparatively new portion of our neural system, which was once called "the silent area" of the brain because its function was largely unknown and no activity was indicated there. Paul MacLean considered the prefrontals a fourth evolutionary system, however, and called them the "angel lobes," attributing to them our "higher human virtues" of love, compassion, empathy, and understanding, as well as our advanced intellectual skills. Antonio Damasio considers prefrontal function the source of all higher intellectual capacities such as our abilities to compute and reason, analyze, think creatively, and so on. Elkhonon Goldberg adds to the growing literature on this subject, showing how elaborate the connecting network is between prefrontals and all other parts of our brain. He points out that the more advanced an evolutionary module or lobe, the more subject it is to damage, and explains how damage to any part of the brain affects the function of the prefrontals.

The ultimate function of the neurons in the prefrontal cortex, according to Patricia Goldman Rakic, ". . . is to excite or inhibit activity in other parts of the brain," which, according to Allan Schore, builds affect-regulation, the lifelong ability to regulate our emotional reactions, control our impulses, or moderate the survival reflexes of our ancient R-system. Evolution's latest addition to the brain was included, it seems, to govern the actions of the earlier modules of our threefold brain, a service that proves critical to a child's development before the secondary stage of prefrontal growth, which opens in mid-adolescence. There is little research available, however, on the nature of this secondary prefrontal growth spurt.

We do know that the prefrontal cortex plays a role in language development, interacting with the temporal lobes, which are involved with all aspects of sound and are located on either side of the left and right hemispheres of the neocortex. Our earliest concrete language of childhood is

---

2. Bernadette Roberts, my friend David Spillane, and other acquaintances have reported on a prolonged period of intense heat and even pain at this area of the forehead either following years of meditation or at critical periods of their spiritual journey. The effect flared up in my life when I was fifty-three. Both Spillane's and my experience of this lasted for years. This phenomenon's locale within what neurophysiology calls the orbito-frontal loop is intriguing.

now thought to involve a dynamic between the prefrontals and the right temporal lobe, while the abstract semantic language that unfolds at about age twelve may involve an interaction of the prefrontals and the left temporal lobe.

In this chapter we will look at what is known about the stages of prefrontal development and identify what remains largely a mystery relative to the function—and even the very existence—of this newest neural structure.

THE STAGES OF PREFRONTAL DEVELOPMENT

▼

Significantly, the prefrontals unfold for development in two stages I refer to as primary (meaning early) and secondary (meaning late). Primary prefrontal growth and usage develops rapidly after birth, in parallel with the rest of our brain. At about age fifteen the majority of the threefold brain completes its development and stabilizes. Only then does the secondary stage of prefrontal growth begin, opening with a major growth spurt, or outpouring of new neural material. Because it was discovered only recently— in the late 1980s—this aspect of development has not yet been acknowledged on a broad academic level.

It is this secondary stage of prefrontal growth, about which we are only beginning to learn, that relates to the two roads we as humans may take— both the path of violence and the path of transcendence hinge on the outcome of this mid-adolescent event.[3]

PRIMARY PREFRONTAL DEVELOPMENT
AND THE ORBITO-FRONTAL LOOP

In the last chapter I mentioned that the development of the human brain in utero during the trimesters of pregnancy mirrors the evolutionary development of our sensory-motor (reptilian) brain, emotional-cognitive (old

---

3. At the same time that the prefrontals have their secondary growth spurt, the ancient cerebellum undergoes corresponding growth. The cerebellum is made of extensions of all three brains in our triune system, and is involved in just about everything we do, though primarily speech and movement. It is made of trillions of granular cells that are quite different from ordinary neural cells. It is noteworthy that granular cells of the same order are also a significant a part of the prefrontal makeup, and that very strong neural links exist between the prefrontal cortex and the cerebellum.

mammalian) brain, and verbal-intellectual brain (neocortex). The primary prefrontals, on the other hand, begin their major growth after a child is born, continuing this development through the first year after birth—with some qualifications: The caregiver's emotional state and the extent of nurturing and care an infant receives can actually affect the development of the prefrontals at the cellular level. From the very beginning, then, the prefrontals are experience-dependent, shaped by the environment the child experiences.

After the first year, the primary prefrontal stage unfolds in parallel with the various systems of the threefold brain that begin to develop in utero and continue to grow from birth, such as sensory-motor capacities, speech, visual and spatial abilities, and so on.

In the earliest period of infancy, for instance, the prefrontal lobes develop parallel to the growth of the sensory-motor system; in the toddler period they develop in tandem with the emotional-cognitive system; in the dreamlike intuitive period of the four- or five-year-old the prefrontals parallel right-hemisphere, temporal lobe development; and in the operational period of the seven- to eleven-year-old, they govern the synthesis of the left and right hemispheres. This pattern of parallel growth between the prefrontals and each of the older brain structures establishes ever-stronger neural connections between the prefrontal cortex and the threefold brain. The end result is a prefrontal system that is thoroughly and dynamically connected to every part of the brain—every module, gland, lobe, hemisphere—which makes for the machinery by which, through the prefrontals themselves, we can regulate and monitor our brain's neural structures.

In their early primary stage, then, the prefrontals unfold not so much from their own inherent capacity for development, as the older brain systems do, but more through their influence on the unfolding of these earlier evolutionary systems. The prefrontal lobes parallel the growth of the other systems because they have an important task at hand: Their main objective at this time is to govern each module or lobe of the threefold brain in its sequential unfolding in such a way that each older system forms according to the needs of the prefrontals in their secondary stage of development during the child's mid-adolescence. The task of the prefrontals is to turn the unruly reptilian brain, old mammalian brain, and neocortex into one civilized mind that it may access later. It is only when this has occurred that the secondary prefrontal stage can unfold as designed.

This ability of the prefrontals to monitor development in the other parts of the brain involves far more than flipping a chemical switch. It requires a sophisticated biological function that gives lower systems the ability to rise to the higher evolutionary level of the prefrontals much as that great, curving serpent rose to break through the top of the head of Egypt's Sphinx.

This function enabling the lower neural systems to move beyond their initial capacity develops at a specific time and in a particular way. A prominent growth spurt in the prefrontals takes place toward the end of the first year of life, at the beginning of the toddler stage. The outcome of this period of specialized neural growth and the capacities garnered by the toddler through it depend, as does all prefrontal development, on the nature and amount of nurturing from the child's primary caregiver, an issue we will pursue in part 2 of this book.

For now, let us assume that nurturing is more than adequate. As a result of this second-year growth spurt, then, rich neural connections are established between the forward-most section of the prefrontal lobes (which have developed in the first year after birth) and the highest region of the emotional (old mammalian) brain, which was established in utero.[4] The name for this link between an ancient brain module and the newest lobe in our head is the *orbito-frontal loop* (because it is centered behind the orbit of the eye), and it is responsible for determining a person's relationship and mental capacities in life.

It is significant that the orbito-frontal loop is established just before the infant gets up on his legs to explore the great world beyond his nest. In this way nature provides the new neural material needed for the emotional imprints, "constructions of knowledge," and relationships the child will create in his second year of life. In fact, Jean Piaget, the Swiss biologist and child

---

4. This region of the emotional-cognitive brain, called the *cyngulate gyrus*, contains, among other elements of its rich heritage, the blueprints for herding instincts—which precede human society—and those species survival instincts that impel us to nurture and protect our offspring. Elkhonon Goldberg thinks that perhaps the cyngulate should be considered part of the neocortex rather than of the limbic system because it seems to be a more advanced evolutionary module than others in the emotional-cognitive brain. At any rate, nature tries to link the two at this point.

psychologist, who coined the term *construction of knowledge,* pointed out that a child's direct sensory-motor exploration is required to build the neural constructions of cognition and learning.

Thus all primary and secondary prefrontal functions having to do with our relationships and our control of R-system instincts (survival, protection, sexual drives, appetites, and so on) center on the orbito-frontal link with the emotional-cognitive brain. And further, the "affective tone" or emotional state experienced by the toddler during the exploratory period after age one determines the nature of the orbito-frontal loop and its ability to function. Allan Schore explains why it is that a toddler's emotional state during this time of world exploration determines whether or not the orbito-frontal connection is established and used or largely lost: He describes how the orbito-frontal linkage is entwined with the care a toddler receives and how this, in turn, determines the lifelong shape and character of that child's worldview, mind-set, sense of self, impulse control, and ability to relate to others. With this explanation in mind, it's impossible to overestimate the importance of the orbito-frontal function.

OTHER STAGES OF BRAIN GROWTH DURING
PRIMARY PREFRONTAL DEVELOPMENT

Beyond the growth that leads to the development of the orbito-frontal loop, a number of other neural growth spurts occur in other parts of our brain at birth, age one, age four, and age seven, with additional shifts of function taking place at ages nine and eleven. These shifts coincide with the developmental stages of childhood called "windows of opportunity," during which particular blueprints for intelligences and abilities unfold relative to the neural modules or parts of the brain ready for development at those times. It seems that different modules are hardwired for different categories of experience at each of these stages, and that the primary prefrontals, in correspondence with specific genetic predispositions, unfold parallel to these stages. This synchronous activity allows the prefrontals to bring about the critical modifications of the capacities within these categories of experience that are needed to pave the way for future prefrontal development. (See figure 2.) With such strong connections to all portions of the brain, it's easy to see, as Goldberg points out, that damage to any part

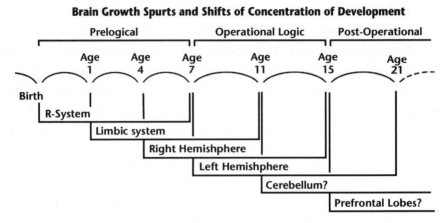

Figure 2. Brain growth spurts and shifts of concentration of development. Illustration courtesy of *Magical Child Matures* (New York: E. P. Dutton, 1985).

of the brain or failure of any part to develop can profoundly affect the prefrontals.[5]

The entire system of brain development can be likened to the encoded blueprints of a house. The neural foundations of a new life must be established first, followed by walls, roof, trimmings, and finally furnishings. As you can see in figure 1 (see page 23), the reptilian brain, with its sensory-motor intelligence, develops first; followed by the old mammalian brain with its capacity to relate intelligently; then the right hemisphere of the neocortex, with its creativity, imagination, and ability to respond intelligently to new and unknown situations; then the left hemisphere, with its control over established learning and routines, and its capacity for analytical logic, abstract thinking, and intellectual power; and then—to reach the point we have come to thus far—the various interactions and syntheses of all these systems.

---

5. Eklhonon Goldberg's argument for a gradiential action in the neocortex, as opposed to the more isolated modular rigidity of the older systems, is an important observation. *Gradiential action*—action that is progressive by logical degrees—implies a far richer network interaction, dynamic exchange, and fluidity among the neural fields of the neocortex than exists among the older, specialty-limited modules of the brain such as the amygdala and hippocampus. We have erred, Goldberg points out, in assuming a modular action in the neocortex like that found in the two older brains. His observation includes a wonderful analogy among social structure, governments, and brain organization. See the conclusion of his book *The Executive Brain* (New York: Oxford University Press, 2001).

In "building" a child's brain and corresponding capacities, each window of opportunity closes as the next one opens in its appropriate evolutionary sequence—though perhaps it's more accurate to say that a developmental stage doesn't close so much as the energy and attention of growth simply shift to the next module or portion of the blueprint waiting to be read and made real. According to this pattern, our neural structures are primed and ready for stimulus and response in their sequential evolutionary order. At each shift from one developmental period to the next, the child's energy and attention also shift, looking for that next window of opportunity.

THE ROLE OF EXPERIENCE IN PREFRONTAL DEVELOPMENT

All of this sequential development takes place in our first fifteen years of life as preparation for the mid-adolescent prefrontal growth spurt that allows us to rise and go beyond the limitations, constraints, and shortcomings of the earlier neural system.

Goldberg makes clear how critically overall development affects the prefrontals as they prepare for this leap, and how much they, in turn, affect overall development. Miss one sequence and the entire structure is at risk. Try putting a roof on a house with no walls, or framing walls with no foundation! To do her part in making certain no step is missed or shortchanged, nature provides us with genetic blueprints, but only as bare outlines for possibility and action. It might be assumed that these genetic blueprints within a child will suffice as the stimuli for neural development. Those genes, however, must be stimulated by the child's interaction with the actual expression of the capacities they imply at the time when those capacities within the child are ready for development. This is why one infant or child can't model for another. Someone who is fully able to do something or behave in a certain way must perform the role of model if a similar ability is to be awakened in that child.

If, however, a child's environment does not furnish the appropriate stimuli needed to activate prefrontal neurons—if the model imperative has not been fulfilled—the prefrontals can't develop as designed. Their cellular growth itself becomes compromised and faulty.

When a child's environment does furnish a model, it follows that the nature of the intelligence or ability awakened and developed in the child will reflect the nature of the child's model. Likewise, though to an indeterminable degree, the nature or extent of governance a child displays over the impulses of the more primitive systems in his or her brain will depend upon the nature of the governance exhibited by the model caregiver.

Thus, with eons of success behind her, nature assumes that her agenda for successful development will be met with appropriate caregiver and environmental response and nurturing. Statistically, at least, such response by parents or caregivers and society has always been forthcoming to some extent or we wouldn't be here. Overall, nature's blueprint unfolds regardless of the success or failure of any individual stage of development. If the window of opportunity offered by a particular stage is missed, she blithely opens her next one on schedule as though all were well. Despite the fact that each stage of development depends on the success of the previous one for its own success, the stages keep coming. Nature's schedule, not our response, is at the controls. She doesn't call a time-out if a child's nurturing environment fails to respond to his needs.

Consider, for example, that new teeth appear in humans about every six years. Though a child's baby teeth might fall out from neglect and bad diet soon after appearing, those six-year molars, twelve-year molars, and wisdom teeth will still appear on schedule, even if they follow the same fate as the child's first teeth. While the sequence of unfolding is statistically stable, more or less, a nurturing environment for the plan is, to say the least, variable, as in all stochastic systems.

INHIBITIONS AND EXCITEMENTS

Elkhonon Goldberg refers to the prefrontal cortex as giving us our civilized mind. At least, we are given the chance to develop civility through the prefrontals, and it is this civility that is a prerequisite for the development of our capacity to transcend, to fill our role in evolution as well as curb our suicidal violence, and to survive.

The prefrontal cortex, as our highest evolutionary brain, acts to lift up our ancient animal instincts to a higher order of their own functioning through the dynamic prefrontal action referred to (rather dispassionately) by Patricia Rakic as *cellular inhibition* or *cellular excitation*. This dynamic is designed to bring our lower survival intelligences (sensory-motor, emotional, and verbal-intellectual) into a proper resonance with the higher prefrontal functions that will unfold in late adolescence. As intent precedes the ability to do and as the new is always built upon the old, nature provides, as antecedent to the appearance of the secondary growth of prefrontals, some fifteen years of catalytic action from the primary prefrontal functions.

SECONDARY PREFRONTAL DEVELOPMENT

The prefrontal growth spurt at about age fifteen signals a second phase of development for which the primary prefrontals, through their governance role, helped prepare. The late-unfolding prefrontals at this mid-adolescent age are the most fragile of all systems in our brain and are critically dependent on an appropriate foundation, which is the entire triune structure of reptilian brain, old-mammalian brain, and neocortex. Thus the full unfolding of this latest prefrontal addition must await the completion of the foundations on which it will stand, and nature works to ensure that these three lower systems will be brought up as proper foundations for her newest and potentially most powerful brain.[6]

Consider, for instance, that the ancient reptilian instinct for species survival plays out in a ritualized, near-mechanical reproductive reflex in the snake or lizard. Through being lifted up into a higher order by each additional module and lobe, including the prefrontals, this instinct evolves in us to such magnificent levels as immortalized in the legends of Tristan and Isolde; Romeo and Juliet; and Helen of Troy, whose face launched a thousand ships; or the erotic, rapturous poetry of the Song of Songs, Rumi, Ibn Arabi, and John of the Cross and his mentor, Saint Theresa, wherein sexuality is lifted to a high transcendent order indeed. Recall Eckhart: When the higher incorporates the lower into its service, it transforms the lower into the nature of the higher. This is the key to nature's plan for our biology. At mid-adolescence the prefrontal cortex incorporates our instincts of the lowly serpent, and they are subsequently transformed into transcendent power.

---

6. Because the secondary stage of prefrontal growth is the highest evolutionary movement within us, it is the most fragile—precisely as the toddler stage is so fragile. This means that the emotional nurturing received at that mid-teenage period serves as a major determinant in the success or failure of this latest opening of intelligence. Therefore, the same qualification made concerning the growth of the orbito-frontal loop at the critical toddler period must be made concerning the period of prefrontal growth in our adolescence, an issue we will touch on later.

How, exactly, is this accomplished? At about age fifteen all parts of the brain other than the prefrontals myelinate, or stabilize, making permanent all developments that have been achieved up to that time.[7] Because function follows form in the pattern of unfolding and development of neural modules, lobes, and hemispheres, as each brain module realizes its essential structural form, full function of its encoded program of possibility begins to unfold and develop, at least to the extent that appropriate stimuli and nurturing are provided.

In this way, during our fifteenth year, when the rest of the brain has ideally matured and is well established, our most recent evolutionary brain, the prefrontals, undergoes its secondary growth spurt. These lobes are busy throughout late adolescence, building their infrastructure ("laying down their neural tracks," as researchers describe it). Although a preliminary peak of the secondary prefrontal growth occurs at about age eighteen, this new neural growth is not completed until about age twenty-one, some six years after the rest of our brain has solidified its form and function.

In looking at the patterns of brain growth as a whole, it is easy to understand why the growth of the secondary stage of prefrontal development must await the maturation and myelination of the triune brain structure. Just as common sense shows that the first year of an infant's life must be devoted to the development of a functional sensory-motor system in order for the child to take in the information needed to develop his emotional-cognitive system, so at each stage of development each new system is dependent upon the full function of the system that developmentally preceded it. In the same way, the secondary stage of prefrontal growth—an extraordinary undertaking—depends upon a fully mature and supportive triune brain structure.

Once growth has been completed, the mature potential of the prefron-

---

7. Myelin is a fatty protein that encases the long axons involved in neural field communications. Forming as a result of the use of those fields, it makes the transmission of signals faster and more economical so that less energy is required in the process. Because myelin is largely impervious to the hormones used to dissolve neural fields in periods of neural pruning (the housecleaning of noncontributing cells), myelination makes permanent the learning, imprints, and developments gained to that point. This includes all of our bad habits, as the tobacco companies realized years ago when this research was published, and they immediately capitalized on it, encouraging youngsters to smoke.

tal cortex should be as dramatically different from what existed before as the formal operational stage of age twelve is from the concrete operational stage of seven, and as both are from the toddler's first excited explorations. In fact, the development of these new prefrontal additions should ideally result in a mind that is so remarkably different from the one we operated with before that it would present to us in full the biological possibility of transcendence.

Of course, each major stage of development offers a window of opportunity distinctly different from anything that has come before. When a higher neural form in our brain completes its growth and begins its full function, a new form of reality and a larger world unfold to us and distinctly new behaviors and abilities fill our repertoire. As we grow from birth to age twenty-one, the strength and complexity of these stages increases exponentially. The fifteen-year-old brain is as different from that of a five-year-old as are the body sizes and behaviors of the two. Logically we could expect that on the completion and maturation of nature's latest, largest, and highest brain at age twenty-one, we would possess capacities more dramatically different from and more powerful than anything previously experienced.

But, in fact, nothing much happens at all.

THE MISSING STAGE

In 1988 Oxford University Press published an intriguing work edited by physicists C. N. Alexander and E. J. Langer titled *Higher Stages of Human Development: Adult Growth Beyond Formal Operations.* Even though this work was compiled before discovery of the late-prefrontal growth spurt, two items stand out in relation to what we now understand about this developmental stage. First, the editors showed how the increase of intelligence at each stage of development is disproportionately greater than the increase exhibited in the previous stage, similar to the order of increase found in the Richter scale for measuring earthquakes. Thus the intelligence increase in the stage designed to open at late adolescence is an order of magnitude vastly beyond that of the previous stage, suggesting an intelligence in no way related to anything coming before. Adding up all experience and knowledge gained to that point gives no hint of the possibilities ready to unfold somewhere around age twenty-one. A similarly massive gain in intelligence can be found when the six- to seven-year-old shifts into operational thought.

Second, Alexander and Langer showed that as each developmental stage offers a more advanced intelligence, a significantly smaller percentage of our populace achieves that stage. The more advanced the intelligence that unfolds through evolutionary process, the fewer the people who develop it. Ken Wilber spoke of a pyramid effect in development: The mass of humans clump at the base of the pyramid, which is the sensory-motor stage, and progressively fewer are found at each higher level, much like the gospel comment of a narrow gate that few find.[8]

Consider, then, that at age twenty-one or so evolution's latest neural structure completes its growth. The most marked change and potential window of opportunity of all should open—but nothing significant happens other than a possible increase in the young adult's ability to reason reflectively, and even this is not a secondary prefrontal development, but rather a maturation of a primary ability.[9] Allan Schore does not take into account Wilber's sharp falling off of numbers in our populace who arrive at each of the higher stages, but he devotes considerable attention to the pathologies of the adolescent and adult who are stuck down at the pyramid's lower levels.

Nature's invitation to rise and go beyond is made over and over, at each stage of our development. But of course development doesn't always unfold as nature designed, to say the least. Indeed, in this random system few of us experience that fortunate synchrony of events that allows her agenda to be carried out without a hitch, which adds the sharp piquant

---

8. Alexander and Langer argued on behalf of Transcendental Meditation being just that, a developmental practice that opens us to our transcendent nature. I let their argument rest with their book without refutation on my part because I make a similar argument on behalf of a historical occurrence pre-dating their book by some two millenia. Both events may be valid and both have been warped and neutralized by culture.

9. Three-year-old children can outperform chimpanzees, our proposed cousins, in reasoning and modification of more instinctive behaviors. This modulating capacity grows with each passing year—or should; the six-year-old can out-reason the three-year-old, the nine-year-old the six-year-old, and so on. Thus a more expansive reasoning at twenty-one—which is the only major developmental change at that age, as noted by developmentalists—is still a primary prefrontal capacity that is observable in rudimentary form at quite a young age.

flavor of risk and the unknown to every aspect of this stochastic life. For most twenty-one-year-olds, when the infrastructure of evolution's newest brain is complete and the neural form is ready for full functioning, nothing seems to happen that is in any way commensurate with the newness, size, and long, drawn-out formation of the complete prefrontal cortex.

On the contrary, evolution's latest neural addition seems to lie largely dormant within us despite the fact that it seems it should offer a discontinuously new potential, a new reality—a whole new mind. The primary prefrontal function, formed in the earliest years, continues but there is little to indicate an evolutionary shift of function and behavior, or revolutionary change in life, as can be rightfully expected.

Unfortunately, there is little or no direct research on this developmental stage. Allen Schore's research into prefrontal-emotional interaction explores some of the later pathologies that take place from failure to nurture the child, but does not address late-adolescent development as an issue. Antonio Damasio makes no reference to this prefrontal growth spurt at mid-adolescence, nor does Elkhonon Goldberg. Biologist Carla Hannaford, in her splendid book *Smart Moves,* does discuss this late prefrontal development, but she is an exception. The phenomenon hasn't yet been widely acknowledged, perhaps because nothing significant happens. But the very fact that nothing significant happens is itself significant.

A TRINITY OF GREAT EXPECTATIONS

By way of concluding a discussion of the prefrontal cortex, which grows to full maturity from ages fifteen to twenty-one, and by way of guiding us into our next exploration, consider three characteristics of adolescence: A poignant and passionate idealism arises in early puberty, followed by an equally passionate expectation in the mid-teens that "something tremendous is supposed to happen" and finally by the teenager's boundless, exuberant belief in "the hidden greatness within me." A teenager often gestures toward his or her heart when speaking of these three sensibilities, for the heart is involved in what should take place. Recall what Eckhart said: There is no being except in a mode of being. The teenager's gesture toward the heart when expressing these great expectations shows that the heart is involved in these feelings and thoughts.

The brain is the heart's modus operandi, or means, for transcendent experience, and nature intends this highest stage to be ready to unfold fully at twenty-one. Development of this new stage would be lifelong if that stage were to unfold. Rudolf Steiner clearly describes these higher stages, pointing to age thirty, for instance, as the time of another step toward transcendence.

Opening to this mature developmental sequence is the adolescent's great expectation. We might think the intelligence of the heart is present all the time and permeates all being, but the heart's latent capacity for deep universal intelligence must, like the brain, be provided with models for its full growth and development. If no nurturing or modeling is given, the powers of the heart can't unfold—they will be dormant for life.

In part 2 we will return to the prefrontals and follow through with some thoughts on the question raised here—why it is that nothing significant occurs when the prefrontals finally and fully mature at age twenty-one. But for now we turn to an overview of the major biological apparatus within us and the seat of our greatest intelligence—the heart.

▼

THREE

THE TRIUNE HEART:

Electromagnetic, Neural, Hormonal
(Universal, Personal, Biological)

There is no being except in a mode of being.

—MEISTER ECKHART

Some five or six decades ago, in some biology class or other, we extracted a cell from a live rodent's heart, put it in vitro, and examined it through a microscope. That lonely cell continued to pulse evenly for some time but then fibrillated (pulsed spasmodically) and died. We could take two live heart cells, keep them separated on the slide, and, when fibrillation began, bring them closer together. At some magical point of spatial proximity they would stop fibrillating and resume their regular pulsing in synchrony with each other—a microscopic heart.

The two cells didn't have to touch for this magical bonding to occur, and could in fact be separated by a tiny barrier. Relationship counts, it seems, but the question was, how did those cells communicate across a spatial and even physical barrier? In the ensuing half-century, through the wonders of scientific research, that mystery has been solved.

ELECTROMAGNETIC BONDS

▼

Electromagnetism is a term covering the entire gamut of most energy known today, from power waves that may give rise to atomic-molecular action to radio waves; microwaves; and infrared, ultraviolet, and visible light waves; from X rays to gamma rays. A heart cell is unique not only in its pulsation but more in that it produces a strong electromagnetic signal that radiates out beyond that cell. Spatial proximity of those two heart cells on the microscope's slide brought their respective electromagnetic fields into conjunction, at which point the two frequencies entrained or meshed in a coherent synchronization. This apparently lifted the disorder of fibrillation leading toward death into the ordered heart rhythm of life. Relationship counts indeed.

All living forms produce an electrical field because in some sense everything has an electromagnetic element or basis, but a heart cell's electrical output is exceptional. That congregation within us, billions of little generators working in unison, produces two and a half watts of electrical energy with each heartbeat at an amplitude forty to sixty times greater than that of brain waves—enough to light a small electric bulb. This energy forms an electromagnetic *field* that radiates out some twelve to fifteen feet beyond our body itself.

We have all heard about brain waves, which, by placing sensitive electrodes on our skull, we are able to read through an electroencephalogram, or EEG. Although the brain as a whole consumes some twenty-five watts of energy when in full action, these individual brain waves recorded on the EEG are minuscule, so sensitive that the tiniest shift of frequency or alteration in the strength of a signal can have a serious impact on brain function. Similar to the way we record those faint brain waves, we can record and read the far stronger signals from the heart through an electrocardiogram, or ECG. We can do this by placing electrodes on our body to pick up the heart's frequency signals, but a reading can also be taken three feet away from the body without any contacting electrodes because the heart's frequency spectrum is quite powerful within that three-foot domain.

THE HEART IS NOT (JUST) A PUMP

The heart's electromagnetic field has a further source of power, an explanation of which requires a bit of a detour. Recent research strongly qualifies the nineteenth-century notion of the heart as a pump. When this analogy was first made, the steam engine was a new invention and the pumping action of its pistons fascinated early physiologists. Certainly the heart maintains a pumping action, but not in the manner long presumed.

Physicists and physicians recently calculated the pressure needed to force liquid through fifteen miles of tubing (the average length of a body's vascular system—not taking into account its thousands of miles of nearly microscopic capillaries). They found the necessary pressure would require a diesel engine with power sufficient to run a Mack truck.

Fortunately, circulation is accomplished not by such power but by the strength of a combination of factors including synchronous contraction-expansion of the blood vessels themselves (arteriosclerosis cripples an artery's capacity to contract and expand); the motility and plasticity of blood cells (they change shape according to the size of the vessel they are moving through); contraction of the skeletal muscles; and the automatic propensity of liquids to move through capillaries. Additionally, new findings show that blood may flow in spiral-like vortices similar to a whirlpool in a river or the swirl of water that gushes down a drain. Grooves in the blood vessels themselves may aid the spin, acting on the blood much like the grooves in the barrel of a rifle act on a bullet.

When the heart closes its valves, the flow of blood is constricted, increasing the pressure throughout the body. At the appropriate pressure maximum, the heart valves open to release the blood into the heart chamber. This rush of swirling blood forms an even stronger vortex as it enters the heart's chambers. A physicist from NASA patiently explained to me that inserting a single ion into such a vortex will create a powerful electromagnetic field. The actual muscular pumping of the heart contributes to the vortex force as the blood moves from chamber to chamber, then on to the lungs and back. It is this muscular action in combination with all of the other factors presented that produces that life-giving beat.

THE TORUS, THE HEART, AND THE UNIVERSE

But this is only half the story. Perhaps you remember as a child dropping iron filings on a piece of paper, holding a magnet beneath, and watching as the filings formed arcs out from and back to the poles of the magnet. A roughly similar action results from the electromagnetic (or em) energy produced by the heart. This em energy arcs out from and curves back to the heart to form a *torus* (see figure 3), or field that extends as far as twelve to fifteen feet from the body. The first three feet are the strongest, with the strength decreasing with distance from the heart according to ordinary physical principles.

The *dipole*, or axis, of this heart torus extends through the length of our body, more or less, from the pelvic floor to the top of the skull. (I recently

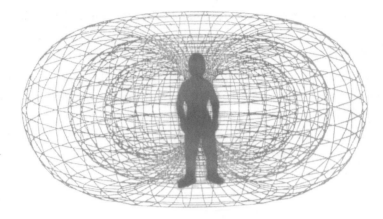

Figure 3. A computer projection of the heart's torus. Courtesy of HeartMath Institute.

learned that the dipole may slant rather than run neatly parallel to the spine.) Early computer representations of the heart's energy field look like a fat, neatly symmetrical doughnut, as in figure 3. Recently, magnificent images of the actual energy arcing out from a living human heart—the torus in its immediate formation—were made at the University of Utah using a new magnetic imaging device. (See figure 4.)

These actual arcs are not the neat symmetries of earlier computer projections, but organic, constantly shifting, living forces. Both the perfect symmetry of the computer rendering and the living asymmetry of the actual phenomenon as captured by magnetic imaging are, to me, nearly sacred,

considering all they represent. This torus function is apparently holographic, meaning that any point within the torus contains the information of the whole field. That is, at any location within the heart's field, no matter how infinitesimal, all

Figure 4. A magnified view of the currents near the heart. Courtesy of the University of Utah Computer Department.

the frequencies of the heart's spectrum are present. Though this characteristic may be puzzling to common sense, it is a matter of great significance to our brain and body. Further, according to physicists, a torus is a very stable form for energy, which, once generated and set in motion, tends to self-perpetuate. Some scientists conjecture that all energy systems from the atomic to the universal level are toroid in form. This leads to the possibility that there is only one universal torus encompassing an infinite number of interacting, holographic tori within its spectrum.

In fact, our earth is the center of such a torus. Earth's magnetic poles are the extremities of the dipole from which the lines of force arc out and around our globe just as the heart's field does around our body. And like the heart's field, the electromagnetic field of the earth is holographic—it can be read in its totality from any single isolated spot on the earth's surface.

Our solar system is apparently toroid in function, with the sun at its center as our heart is at our center. Fluctuations in the energy fields of the sun produce disturbances in the corresponding magnetic lines of earth, such as those that result in the aurora borealis, or northern lights. We seem to live in a nested hierarchy of toroid energy systems that extend possibly from the minuscule atom to human to planet, solar system, and, ultimately, galaxy.

Because electromagnetic torus fields are holographic, it is probable that the sum total of our universe might be present within the frequency spectrum of any single torus. As it turns out, then, William Blake, able to see a world in a grain of sand, demonstrated insight as well as creative vision. An individual torus may participate holographically within a universal torus over a wide range, from the simple wave particle to the incredible complexities of our heart, brain, and body, all of which are electromagnetic by nature. One implication of this is that each of us centered within our heart torus is as much the center of the universe as any other creature or point, with equal access to all that exists.

In the nonlocality of a frequency realm, the most minute microportion conceivable would still contain the information of the whole. Years ago, neuroscientist Karl Pribram proposed that the brain draws its materials for constructing our world experience from a "frequency realm that is not in time space." So the hierarchy of frequencies in which we move and have our being would carry, in some manner, the information out of which our brain and body build our lived experience.

Indeed, we live in fields within fields of a holographic electromagnetic display where all information is somehow present within every minute part of any particular frequency. Each part is thus representative of the whole, with our human heart somehow the genesis of our personal yet uniquely shared living world. As neuroscience keeps pointing out, no claim can be made of a world unto itself; we are always speaking, by default, of the world presented to us through our brain and body's neural system. The world my brain and body give me is approximately the same as that which you and others are given because our similar physiology draws on the same nonlocalized frequency fields. The slight variations in our worlds—which often lead to mayhem—are due to the minor variations in our neural systems that, in turn, arise from the unique history and development of each of us. The universal-yet-personal nature of the frequency realm of the heart, the potentials of which we all share, expresses again how creator and created give rise to each other.

THE HEART COMMUNICATES

Our heart maintains an intricate dialogue with our brain, body, and world at large and selects from the hierarchy of em fields the information appropriate to our particular experience. The heart also translates back into that hierarchy of fields our individual response to the reality we experience. This dynamic feedback influences and modifies the very fields of energy from which we spring. We enter into as well as draw from these fields, which are apparently aggregates or resonant groupings of information and/or intelligence.

Our emotional response to our world experience thus changes the nature of the materials we draw on for translation into our world experience. How extensive this reciprocation or mirroring might be is indeterminable, for it is stochastic—subject to random chance—as in all dynamic systems.

When brain and heart frequencies entrain, they enter a synchronous, resonant, or coherent wave pattern. (See figure 5.) Though rare in adults, such entrainment is critical to full development of our human nature, and new research is revealing how this can be achieved. In the example opening this chapter, entrainment between the two heart cells lifted them from chaos into order. The same entrainment of heart frequencies occurs between mother and infant during breast-feeding and other close body contact.

Head-Heart Entrainment

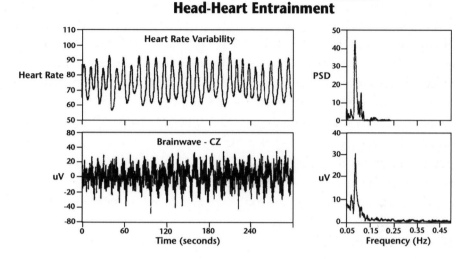

Figure 5. Here the overall average of wave forms produced by the heart and the brain are compared and shown to be synchronous following the employment of the HeartMath "mind tool" called FreezeFrame, a six-step maneuver that brings heart and brain into synchrony during a stressful event, thereby reducing stress and opening new paths of response. Such synchrony occurs within a short period of time and indicates a corresponding shift from hindbrain to forebrain, with energy-attention centering in the higher creative intelligence of the prefrontals. Courtesy of HeartMath Institute.

Through this dynamic, the mother's developed heart furnishes the model frequencies that the infant's heart must have for its own development in the critical first months after birth. In a state of full frequency match, the body, brain, and heart produce a single coherent frequency pulse or wave form, and a similar resonance occurs between infant and mother. (See figure 6.)

Although the frequency spectrum of the heart covers only a 30- to 40-H_z span, indications are that through any single one of these heart frequencies an indeterminable amount of information or material for experience can be conveyed. Consider how a single radio or television band (such as 1500 on your radio dial) can convey a great deal of information. (See figure 7.) The heart's field doesn't need to contain the full spectrum of frequencies available to us, however. Our physical experience may result from frequencies of earth, sun, and beyond, a larger frequency realm within which the heart torus is hierarchically nested.

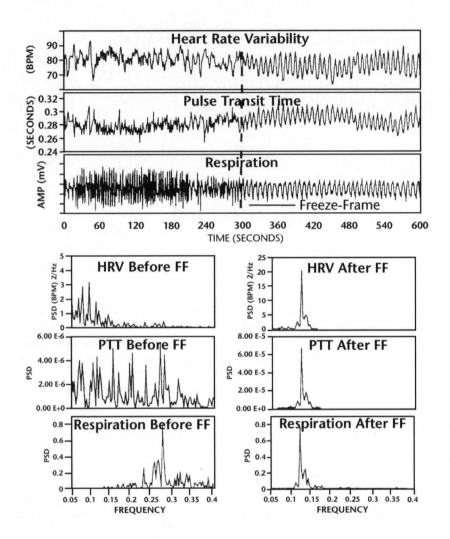

Figure 6. The top graphs show an individual's heart rate variability, pulse transit time, and respiration patterns for 10 minutes. At the 300-second mark, the individual FreezeFramed and all three systems came into entrainment, meaning the patterns are harmonious instead of out-of-sync. The bottom graphs show the spectrum analysis view of the same data. The graphs on the left are the spectral analysis before FreezeFraming. Notice how each pattern looks quite different from the others. The graphs on the right show how all three systems are entrained at the same frequency after FreezeFraming. Courtesy of HeartMath Institute.

ECG Frequency Spectra:

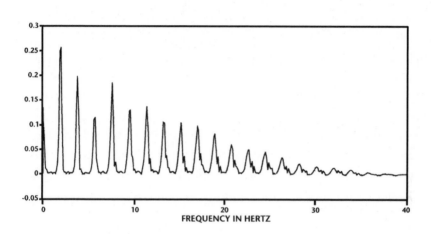

Figure 7. The diagram shows the frequency spectrum, or range of frequencies, in a single instant's radiation from the heart. It is not a time-line graph, but rather an instantaneous one. The height of the wave indicates the amplitude, the distance across the Hertz value. In the holographic frequency range of the heart, any single incidence or sample of wave contains this whole spectrum. Courtesy of HeartMath Institute.

In the 1970s, biologist Bruce Lipton took part in embryonic research that revealed minute electromagnetic fields forming around embryonic forms, whether seeds or embryos within eggs, as soon as conception or germination occurred. This field apparently arises from within the embryonic tissue, curving out from and returning to it much as the heart's em field arises from and returns to the heart, or earth's em field arises from and returns to earth.

It is not surprising that em fields can act on our physical health both positively and negatively on a cellular level. Medical research has found that certain em frequencies can block production of melatonin or prompt a cell to produce enzymes that in turn produce cascades of destructive effects within the cell. On the other hand, physicians in Holland use em fields to heal cancer. DNA is em-sensitive, allowing some em fields to regulate DNA, RNA, and protein synthesis, as well as to induce cell differentiation and morphogenesis. Rupert Sheldrake's morphogenetic fields may involve electromagnetic energy as well as nonlocal effects beyond current knowledge.

THE BRAIN IN THE HEART

▼

Producing electromagnetic energy, then, is the first of our heart's triple characteristics. Neurocardiology, the new medical field exploring the brain in the heart, portrays the second. While neural counting is often controversial, estimates are that half or more of the cells of the heart are neural cells like those making up our brain. Some reports claim 60 to 65 percent of heart cells are neurons, all of which cluster in ganglia, small neural groupings connected through the same type of axon-dendrites forming the neural fields of our brain.

The same neurotransmitters that function in the brain also function in heart ganglia. Through long axons, the spinal cord, and the peripheral nervous system, one aggregate of the heart's neural ganglia connects to the myriad smaller neural ganglia found scattered throughout the body's tissues, muscle spindles, organs, and so on. (In the nineteenth century, Bichat referred to these repositories of neural ganglia throughout the body as "little brains of our animal life," and believed them to be independent from the human brain in our head. Oddly, he may have been half correct.)

Other aggregates of the heart's neural structures have unmediated neural connections with the emotional-cognitive, or limbic, brain, "unmediated" meaning here that no ganglia interrupt or interpret the communication between the heart and emotional-cognitive brain in the way that they do between the heart and the other bodily organs and muscles. (There are no "little brains of our animal life" in this head-to-heart communication!) An ongoing dialogue takes place between the heart and brain through these direct neural connections.

THE HEART IS BOTH UNIVERSAL AND INDIVIDUAL

Years ago my meditation teacher in India, Gurumayi, pointed out that there is essentially only one heart, a universal function expressing through each of us in infinitely varied yet similar ways. This one heart is the universal in our life, and our life is that heart's diverse expression. Perhaps this is one reason that heart transplants are feasible to a certain degree. While critical neural connections are lost in the transplanting process and can't be retrieved, other traits may be picked up that we haven't bargained for, such as some personal idiosyncrasies of the donor. Any heart may take on the color or characteristics of the unique individual in which it functions. Thus, following a transplant, heart transplant recipients occasionally begin to express new characteristics similar to those of the deceased donor. (Lest we get too carried away by this observation, consider that the same phenomenon occurs occasionally in transplants of other organs, such as kidneys. Our new biology points out the continuity and coherence throughout all the spectrums of energy in a being—each part contains, at some level, all the information of the whole.)

The heart takes on the subtle individual colors of a person without losing its essential universality, however. It seems to mediate between our individual self and a universal process while being representative of that universal process. The dynamic between the universal in the heart and the specific variant in the head is, of course, the recurring theme in this book. Meister Eckhart made the audacious claim that when "God becomes Eckhart, Eckhart becomes God." The Sufi Ibn Arabi claimed that God takes on the coloration of each of the infinitely unfolding personalities and is simultaneously each, without any change to God-as-God. Thus, Ibn Arabi pointed out, each name is a name for God, each object is a "face" of God. My meditation teacher Muktananda's favorite phrase was "God dwells within you as you." His teacher, Nityananda, spoke of "God dwelling in the cave of the heart," and urged us to "go there and roam with him." These quaint sayings are taking on the new light of biological support and may have to be considered a bit more seriously.

Once again recall Eckhart's observation that there is no being except through a mode of being. For a universal torus of potential energy-consciousness to be actualized or made real, there must be a mode of and means for such actualization. For us, the heart is the primary mode of being and all else in our life springs from it. The end result is a heart that generates species-specific characteristics that we all share, yet reflects the personal characteristics of each brain and its experiences to make for a unique expression of a shared form. The interaction of these two aspects—the universal and the individual, unity and diversity, creator and created—results in a human being that is the same as all others but different too, and a world that may be as different as that of the Australian Aborigine and technological man.

A step-by-step explanation of the biology of this process of creation of individual-yet-universal man might look something like this:

1. Our brain, with its ganglia extensions throughout the body, is, figuratively speaking, an instrument of the heart.

2. Our heart, in turn, is an instrument or representative of the universal function of life itself.

3. Our brain and body are manifestations of that universal heart's diversity, or individual expression. Brain and body are fashioned to translate from the heart's frequency field the information for building our unique, individual world experience.

   The brain and body then respond to the resulting perceptual experience and determine or interpret its quality. This qualitative analysis, or emotion, is relayed back to the heart, moment by moment. This influences the heart's own neural field, which responds to the emotional report and relays it to the fields of its origin, subsequently changing those fields, if only on a minuscule level.

4. In response to the brain's reports, the heart also changes its own neural and hormonal signals to the body and the brain, and to the production of that em field of information itself. This changed neural, hormonal, and em action then influences the kind of world we experience.

   We live in an environment of feedback or "mirroring" in which creator and created give rise to each other both within us and outside of us.

THE DIALOGUE BETWEEN BRAIN AND HEART

Over many years of research under grants from the National Institutes of Health, John and Beatrice Lacey traced the neurological connections between the brain and the heart. Their discovery of these connections and the ongoing heart-brain dialogue was largely ignored by academic science. Today the new field of neurocardiology has verified and validated the Laceys' work, which means that in time acceptance will follow.

The heart certainly has an intelligence, though this calls for a new definition of the word to differentiate it from cerebral intellect. The heart's intelligence is not verbal or linear or digital, as is the intellect in our head, but rather is a holistic capability that responds in the interest of well-being and continuity, sending to the brain's emotional system an intuitive prompt for appropriate behavior. Intellect, however, can function independently from the heart—that is, without intelligence—and can take over the circuitry and block our heart's more subtle signals.

To better understand the brain in our heart and the concept of an intelligence or wisdom of the heart, we need to understand a bit about the nature of glial cells, which accompany neurons and are as important to the brain as they are to the heart.

THE GLIAL CELLS: SMART GLUE

Although glial cells make up 80 percent of the mass of our brain, very little is yet known about them. (Most brain research thus far has centered on the more accessible neural system.) The word *glia* comes from the Latin word meaning "glue," which is what these tiny cells were long considered to be—glue that held the neurons in place. Eighty percent, however, seems an expensive outlay for a glue that plays no other role. We have since discovered that the glia provide many important services.

Glia are electromagnetically sensitive and form an interactive em field in the brain over and above the electrochemical fields of neurons, with some ten or more glia clustered around each neuron. There is a strong probability that these em-sensitive glia selectively draw from the hierarchy of em fields surrounding us and translate these em frequencies into electrochemical signals available to those neurons, thereby furnishing the information from which our neural system builds our world experience. Gap junctions between glia and neurons provide for a flow of signal-producing calcium ions, and such calcium exchange takes place among the glial cells as well. Glia, then, may both draw selectively on the hierarchy of torus fields and feed back into those fields our neural response—yet another reciprocal dynamic.

Now, extensive selectivity must be made from em fields for any particular life to form and develop. In human embryonic and fetal development, formation of a rudimentary heart comes first, followed next by formation of the brain and, finally, of the body. Long before it becomes our four-chambered heart, our rudimentary heart furnishes the electromagnetic field that surrounds the embryo from its beginning. This em field is, in turn, surrounded by the mother's more powerful heart field, which, in the infant's first months of life, stabilizes the infant's heart field as his heart imprints the mother's. This is another example of nature's model imperative.

Collectively, the glial cells are another player in our translation of information from a frequency realm into our concrete experience, a process that takes place through every cell in our body—including the hundred billion neurons in our brain, their counterparts in our heart, and the glial cells themselves that accompany these neurons. From all these dynamic interactions feeding into one another in resonant patterning, somehow the final perceptual action takes place that gives us our world.

THE HEART AND DOMINION

Consider, then, these three established ways by which the heart regulates and influences brain-body systems. (Sound waves may prove a fourth influence, as Schwartz and Russek suggest.) First and most spectacular of these three—and most unrecognized as yet—is the electromagnetic activity we've already discussed. Heart radiation saturates every cell, DNA molecule, glia, and so on, and helps determine their function and destiny. From this viewpoint the heart seems a frequency generator, creating the fields of information out of which we build our experience of ourselves and the world.

Second, our neuron-laden heart has a myriad of neural connections with the body and direct, unmediated neural connections with the emotional structure of the brain (from which comes the new and popular subject of "emotional intelligence," our brain's translation of that nonverbal, gestalt type of knowing: heart intelligence). These neural connections allow an ongoing dynamic interchange to take place between the brain and the heart quite beneath our awareness. Heart intelligence is not anything of which we are aware, though we surely are aware of the results of these neural interactions. I am reminded of James Carse's wonderful essay "The Silence of God" and Gurumayi's comment that "the language of the heart is silence."

The third level of influence of the heart on the brain is hormonal, which led to the observation of "The Heart as an Endocrine Gland." This was the title from the cover story of *Scientific American* concerning the French physicians Roget et al., who discovered that the atrium area of the heart produces a hormone, labeled ANF, which can modulate and influence the functions of the emotional-cognitive system.[1]

Other heart-generated hormones have since come to light, such as tranquilizers that attempt to keep us in balance and harmony with earth and each other. The heart increases these tranquilizing hormones during pregnancy, giving expectant mothers that peaceful glow that comes from seeing the world through rose-colored glasses, which in this case are truly furnished by nature!

We do well to remember that in utero the heart, as vehicle for the frequencies of our potential world experience, forms first; the brain, as a vehicle for our awareness of those frequency translations, forms second. The heart is intimately connected with every facet of the body and brain through its own neural extensions. As a result, the intelligence of the heart, not the intellect of the brain, has to contend with those signals from the pancreas, liver, spleen, and so on, to maintain order in the ranks. This leaves me, up here in my head, free to dream of dancing nymphs; or to growl about the foolish ideas of preachers, lawyers, politicians, or books by other authors.

A large segment of the earth's electromagnetic spectrum is its radio band, and this may be true of the heart as well. The heart, earth, and sun furnish us the fundamental materials for our reality-making. The heart's em field, like that of the earth, shields us against inappropriate frequencies as best it can, and selects from the larger hologram in which we are nested those frequency groupings appropriate to our growth, development, and ongoing life. In this way what is broad, generic, and universal is expressed on an intimate, personal level, making each of us, even "the least of these our brothers," equal expressions of the totality. Thus, if we have the vision,

---

1. Through this connection, ANF plays a prominent role in every hormonal action of our body and immune system. It also plays a major role in the balance of the sympathetic and parasympathetic nervous systems. (In brief, the sympathetic nervous system is our vigilant guardian, the parasympathetic our nurturing caretaker.) In its positive phase, the sympathetic system speeds us up, as for action; the parasympathetic slows us down, as for meditation or contemplation. In negative states of fear and anger, the sympathetic can produce high levels of cortisol and speed up the system for defense, attack, and rage. In states of fear and anger the parasympathetic can, through the same cortisol, shut us down to a minimum of action, as in depression, ennui, or the paralysis of desensitized withdrawal. In an ideal norm, monitored by the heart, sympathetic and parasympathetic maintain a balanced composure, cortisol-free.

we will see all things as "holy" or whole, as William Blake did, or "see God in each other" as Muktananda did, or find God in the "least of these our brothers," as Jesus did. Anything that is *holographic* is just that, a graph of the whole—and each of us is that.

Because of, or through, this holographic hierarchy of em fields, our heart and brain frequencies can entrain to modulate earth's frequencies to an unknown extent because what we perceive as earth frequencies are translated through our reptilian hindbrain and its sensory-motor system, and the higher can transform the nature of the lower. This accounts for the concrete operational thinking that opens for development around our seventh year. According to Piaget, in concrete operational thinking thought can "operate" on matter, or sensory signals, and change it according to the intent of the thought. This offers us a dominion over our world that we have not as yet accepted or exercised, but which our great model demonstrated for us.

Power is relative in the creative process. A minute shift of frequency in our brain can result in a serious shift in our experience of the world out there. At some point on this spectrum, our experience and the world we experience are reciprocal. The dynamic of creator and created is a play of power and strength we participate in by default and a play we might try to become more aware of so that we can be more conscious of what we are doing.

The following, then, is a sketchy summary of the hypotheses regarding the heart-brain-body dynamic:

1. The heart's electromagnetic field is holographic and draws selectively on the frequencies of the world, our solar system, and whatever is beyond.

2. Through glial action, our neural system selectively draws the materials needed for world-structuring from the electromagnetic fields as coordinated by and through the heart.

3. Our emotional-cognitive brain makes moment-by-moment qualitative evaluations of our experience of the resulting world structure, some of which we initiate in our high cortical areas and others of which form automatically and instinctually in the old mammalian brain.

4. Our emotional-cognitive brain has direct, unmediated neural connections with the heart. Through these neural connections the positive and negative signals of our response to our present moment are sent to the heart moment by moment.

5. The heart's neural system has no structures for perceiving or analyzing the context, nature, details, or logic of our emotional reports. Thus, the heart can't judge the validity of or reason for these reports and responds to them as basic facts. The heart responds on all levels: electromagnetically, through the unmediated neural connections to the limbic brain, and through neural connections to a myriad of body functions. Additional responses include hormonal shifts between the heart and the body and the heart and the brain, and perhaps shifts on sound and thermal levels as suggested by Schwartz and Russek.

In response to a negative signal, the frequency realm of the heart drops from coherent to incoherent. This is a survival maneuver that opens the heart spectrum to an indefinite or variable state. In this fluid situation our body, brain, and heart can respond in new ways to an emergency, if the old survival responses initiated by our lower brain systems are insufficient. (See figure 8.)

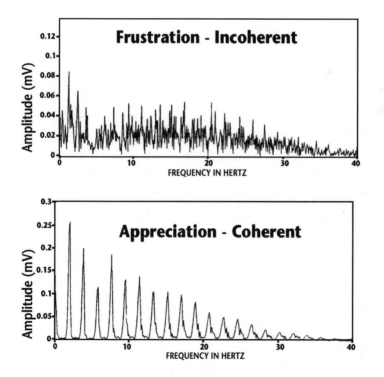

Figure 8. In periods of frustration, fear, or anger, the em spectrum is incoherent. In times when love or appreciation is experienced, it is coherent. Courtesy of HeartMath Institute.

6. When the heart makes such an adaptive shift, suspending its stable norm, our perception changes accordingly. The world we see and experience in a state of fear, rage, dire emergency, competition, or struggle is quite different from that which we experience in a state of harmony and love.

7. During an initially negative response, our brain shifts from the slower reflective intellect of the frontal lobes and neocortex to the quickly reflexive reptilian brain and its links with the emotional-cognitive brain's survival memories and maneuvers. This shift from forebrain to hindbrain is not voluntary or within our awareness—it just happens and always appears as logical, practical, common sense.

8. The dialogue between our heart and brain is an interactive dynamic where each pole of our experience, heart and brain, gives rise to and shapes the other to an indeterminable extent. No cause-effect relationship can be implied in such an organic, stochastic, and infinitely contingent process. This mirroring is another vital example of the creator-created dynamic.

EVOLUTION AND DEVOLUTION AGAIN

In most instances of survival issues the Interpreter Mode in our forebrain aligns with and locks in on these concerns to form a nearly unbreakable tape loop between the forebrain and hindbrain. This shifts all attention and energy to the hindbrain and its survival mode. The forebrain's creative intellect then serves these survival needs while the heart continues to receive variations of the same negative report repeated ad nauseam by the forebrain-hindbrain loop. This response can continue long after the original negative event, so that we live in a closed circuit of resentments and offenses against our neighbor or world, a self-fulfilling and self-perpetuating negative loop.

Again, the heart has no neural structures by which to judge the illusory or delusory nature of the reports the brain sends it. Whether a report comes from the hindbrain and its ancient sensory-motor and survival systems or from the forebrain's creative imagination, all the heart can do is respond accordingly.

As mentioned in the summary above, in a negative state our energy shifts from the forebrain to the hindbrain and, reinforced by the heart's automatic response to negativity, our thinking, judgment, and perception are altered in the interest of defense. In our archaic, defensive mode of mind we have limited our access to our intellectual capacities except as selectively used on behalf of our defenses or revenge strategies. We then use the same selective process to intellectually justify our reflexive action: Defensiveness is always rationally cloaked as common sense, while revenge is always cloaked as justice. These ancient defenses are comfortably familiar; they are locked into our very body and memory, patterns of reaction that make up the commonly shared cultural world in which we participate. Defensiveness is supported by powerful and ancient field energies that are part of our consensus reality itself. In a negative, belligerent state of mind I have crowd comfort, herd support. This is why it is so hard to recognize and alter our defensiveness in those instances where it hurts instead of helps.

Nature's economical habit of building new evolutionary structures on the foundations of older ones has led to our current magnificent potential and terrifying dilemma. Our potential can't be utilized and our dilemma can't be resolved by either intellect or moral and ethical effort alone (if at all). But we have within us this other link, the three-way connection among our emotional-cognitive brain, our prefrontal lobes, and our heart. Here in this connection lies our hope and transcendence—if we can break from the madding crowd. Through understanding and using our heart's intelligence along with our brain's intellect we can resolve our dilemma. Whatever language or rationale it might take, our task is to discover—or rediscover—these two potentials, align them, and come into transcendent dominion over our life.

FOUR

▼

FIELDS WITHIN FIELDS:
Of Frequencies and Neurons

My heart can take on any form, a meadow for gazelles,
a cloister for monks.

—Ibn Arabi, A.D. 1165–1240

Our heart participates in electromagnetic fields within fields nested in hierarchies that are holographic, the whole existing within any part, and all functioning as an integrated dynamic. Mae Wan Ho, Ph.D., a reader in biology at the Open University in England, studies the coherence inherent in each living creature. ". . . [B]ased on empirical findings from our own laboratory, as well as from established laboratories around the world," she writes, "the most suggestive evidence for the coherence of the organism is our discovery, in 1992, that all living organisms are liquid crystalline."[1]

Coherence, in this context, refers to the fact that the trillions of cells and the myriad parts comprising them function together as a unit to produce the mysterious, unified, and magnificent whole called *me.* I must remind myself, as a layperson, and my reader as well, that by this word *organism* biologist Mae Wan Ho means me, this person sitting here at this keyboard, and you, there reading, and that we are not just specimens of research material on the microscope's slide but are, in fact, what all this research is about.

Mae Wan Ho continues:

In the breathtaking color images we generated, one can see that the activities of the organism are fully coordinated in a continuum from the

---

1. Mae Wan Ho, "The Entangled Universe," *Yes! A Journal of Positive Futures,* Spring 2000.

macroscopic to the molecular. The organism is coherent beyond our wildest dreams. Every part is in communication with every other part through a dynamic, tunable, responsive, liquid crystalline medium that pervades the whole body, from organs and tissues to the interior of every cell. Liquid crystallinity gives organisms their characteristic flexibility, exquisite sensitivity and responsiveness, thus optimizing the rapid intercommunication that enables the organism to function as a coherent whole.

When coherent, this polyglot of flesh, blood, sweat, and tears has dominion over the word, naming animals and stars, the parts of our own innards and atoms, and is the source of poetry and song.

In keeping with the ideas presented in part 1 of this book, Mae Wan Ho observes:

> The visible body just happens to be where the wave function of the organism is most dense. Invisible quantum waves are spreading out from each of us and permeating into all other organisms. At the same time, each of us has the waves of every other organism entangled within our own make-up. . . . We are participants in the creation drama that is constantly unfolding. We are constantly co-creating and re-creating ourselves and other organisms in the universe, shaping our common futures, making our dreams come true, and realizing our potentials and ideals.

The field of biology has been changing for quite a while, though the entrenched powers maintaining and controlling the science of biology have kept these changes on the periphery. Lynn Margulis, holding the chair of Distinguished Professor of Botany at the University of Massachusetts at Amherst, made this observation, published in 1988:

> More and more, like the monasteries of the Middle Ages, today's universities and professional societies guard their knowledge. Collusively, the university biology curriculum, the textbook publishers, the National Science Foundation, review committees, the graduate record examiners, and the various microbiological, evolutionary, and zoological societies map out domains of the known and the knowable; they distinguish required from forbidden knowledge, subtly punishing the trespassers with rejection and oblivion; they award the faithful liturgists by granting degrees and dispersing funds and fellowships. Universities and academies, well

within the boundaries of given disciplines . . . , determine who is permitted to know and just what it is that he or she may know.[2]

Mae Wan Ho and Lynn Margulis represent a growing wave of new biologists who break beyond the boundaries to lift the field of biology into new dimensions. This is a growth the guardians of the old boundaries resist, as is typical of all institutions.

FIELDS OF POTENTIAL

▼

Use of the word *field* indicates a habit of our mind to group disparate events, to create from them some unity, a mental category, a taxonomy that lifts the apparent chaos of nature into the order of our thinking. There is a medical field, legal field, a field of education, a field of knowledge, the orbital or wave field of a particle, a field of potential energy, neural fields of brain, fields of stars. All mental, physical, emotional, psychic, religious, spiritual, non-ordinary, or ordinary experience originates as and/or brings about a field. Fields as artifacts of memory or repositories of experience become sources of potential, creations of thought by which we explain our creations to ourselves, or dream up new creations. No field could be bound into a finite system, yet our intellect, once it creates a field, continually tries to establish final boundaries for it. As Rudolf Steiner and poet-philosopher Goethe pointed out, the human mind, as a field effect, is unbounded. There are no limitations to what thought can do, where it can lead, and we ceaselessly strive to discover the full dimensions of self and define that infinite mind. For instance, our many definitions of God and any explanations we offer concerning him (or her, or it) are, by and large, projections we make as a result of our longing to bring closure to this open-ended procedure. "The Son of Man has no place to lay his head," our great model observed, a point ignored by that force of mind that longs to bring all things to closure.

The potential of a field manifests or expresses—presses outward—only through a corresponding neural field (or fields) in our brain. A neural field is an aggregate of neurons linked in such a way that it can translate an aggregate of particular frequencies and present us with a valid perception

---

2. Lynn Margulis, "Big Trouble in Biology" in *Doing Science,* John Brockman, ed. (New York: Prentice Hall, 1988), 213.

or event. Neural fields are worthless without fields of potential frequencies that they can translate into experience, and potential frequencies are worthless without neural fields to translate them into being. These two field categories give rise to each other—another creator-created dynamic—though which comes first will never be determined.

It's important to note that fields don't exist as entities, except as ideas of mind. A "field of potential" exists only as a dynamic interaction with a neural field in our head—*dynamic* here meaning that the action moves both ways, from field to field, from potential to perception and response of the perceiver, then back to the field.

A neuron is made of matter, however flimsy, and matter is an aggregate of particular frequencies, no matter how elusive. Frequencies aggregate in a way that translates other frequencies into a resonant response that is even more elusive and inexplicable: consciousness itself. Physicist David Bohm spoke of consciousness expressing itself as matter and/or energy.

The meaning of *consciousness* has been up for grabs for a long time and remains just as elusive as that of *mind. Brain,* and so *brain-mind,* is more tangible. With some hundred billion neurons, each of which can connect with upwards of a hundred thousand others in mixtures of field interactions, and a trillion or more glia pumping frequencies to be translated, there are no limits to our brain's capacity to create experience. And if you recall, there are no limits to the fields available for these constructions because the same brain can create fields to be translated. We really do reap as we sow, whether or not we are aware of it, though much sowing takes place beneath our awareness.

Fields of information and intelligence are built up through activities among people, and any number of people can take part in these frequency fields. Occasionally the same discovery is made in the same field—say, mathematics, chemistry, or mechanics—by two individuals on opposite sides of the globe who know nothing of each other. Mae Wan Ho refers to this continuum in her own language and brings into being liquid crystals, invisible quantum wave functions, and other hitherto unknowns. None of us knows precisely what we are talking about, but we are compelled to try to describe and explain to ourselves the magic of our minds.

DOES A FIELD EXIST?

▼

No one will ever observe a field as itself because, as already mentioned, fields don't exist as entities. *Existence* comes from the Latin meaning "to be set apart from." A field as itself can't set apart from itself and so can't exist. A field's potential can be set apart, however, and this happens with each discrete, selective neural action relating to and translating an event from that field. This means that a field is quasi-universal: A field of medicine doesn't exist, but doctors do. The medical field is the universal while the doctor is its particular diverse expression. The field exists as that doctor, but the doctor is not the field, which doesn't exist. I give selected fields their selective existence and they give me mine; we give rise to each other. And while fields fall into the creator side of the polarity of our being, that which is created—you and I—can, in turn, create fields. Remember that all dynamics are two-way streets.

In holographic photography any fragment of the holographic plate theoretically contains the entire hologram, but the smaller the fragment, the less clarity the image has. Even so, those with clear vision can see: Blake saw a world in a grain of sand, Muktananda saw God in everyone, and Jesus, who pointed out that "no man has seen the Father," said, "[H]e who sees me sees the Father." There is no difference in essence between part and whole, but part is not whole—though it is all of the whole that can be.

Mathematics, music, language, spatial knowledge, and so on, all outlined by Howard Gardener as multiple intelligences, are powerful fields with endless subfields. Gardener's list is hardly the sum of all field effects (intelligences) we can access because all thought, perception, conception, creativity, discovery, and personal memories relate to, from, or through such field effects. Fields aggregate with their own kind, like birds of a feather.

A new physics speaks of a potential brought into actuality by observation. The observer and that which is observed are participants in a reciprocal dynamic that makes a field neither existent nor able to be perceived—though its realized potential does exist and can be seen. The phoneme pool, or field of forty-two phonemes, though a universal underlying all language, exists only as it is called into play through language or speech itself. The mother speaks and a muscle in the fetus's body responds by moving. The phoneme pool of potential is being expressed by the mother and model and the potential is being activated or given form in a new brain-mind-body network. The language field is not in the brain nor in some hypothetical ether or em spectrum or gene or mind of God. The field exists in its use, in its employment, in the creative action of speech or thought.

The field exists, then, only as parts of itself are set apart from itself. No amount of use or setting apart ever diminishes a field, for it does not exist as an entity but only as a function. Consider the imaginary set of the infinite series of numbers: You can subtract from it vast volumes of numbers forever without changing the nature of the series, which remains infinite and intact. No setting apart, subtraction from, or expressing ever diminishes a field, though no field exists without such setting apart.

Kierkegaard's ecstatic prayer cries out: "[E]ven the fall of a sparrow moves Thee, but nothing changes Thee." The creator is within and intimately involved with every aspect of the created, but is neither the sum total of all that's created nor changed by any aspect of the created. Though the least movement in my heart toward God moves God toward my heart, only my heart-as-God changes—not God-as-God.

Bernadette Roberts wrote of an extraverted, or wide-awake, mystical experience in the Sierra Nevada characterized by her perception of an immense, mind-stopping, and overwhelming intelligence permeating all nature and the universe. Such mystical experience is a fusion of an individual's matrix field with the universal matrix. Occurring in a state of full wakefulness, the individual perceives both his or her field and universal fields simultaneously. The mystic experiences the cusp between being and nonbeing.

Suzanne Segal wrote of her fusion with "the vastness" and her discovery that the vastness perceived its universe through her own sensory system, which was at that point the sensory system of the vastness itself. Segal essentially perceived the universe perceiving itself, but without her, that perception did not exist.

DO ACTIONS STRENGTHEN A FIELD?

▼

Fields seem to grow in strength through the dynamics involved. As in our personal memory, any expression of a field effect enhances and enlarges that field—the more use and feedback a memory receives, the more feedback it generates and the more easily it is accessed. "For to him who has, more will be given"—and while the nature of that "more" is incidental, its function is vital. Jesus fully expected to set up a field effect that any follower of his way both would be strengthened by and would strengthen: "I go to prepare a place for you." His "place" is the creator-created function. Through this dynamic we could all sow our small breeze and reap his whirlwind.

Thom Hartmann wrote that his mentor and teacher, Gottfried Muller, urged him to look for secret acts of compassion he could perform, acts that would be personal, private, and unobserved. An act of compassion done publicly is done for public acclaim, no matter how subtly or covertly that acclaim is accepted by the actor. Further, this public acclaim is the actor's reward. Performed in absolute secrecy for the sake of compassion itself, a private compassionate act strengthens a field of compassion—a field that is sadly absent from our planet. Jesus contrasted the widow's mite, given in secrecy, with the alms-giving of the powerful Pharisee who was preceded by trumpeters on his way to temple so that his magnificent gesture of largesse and virtue might be observed by all. "He has his reward," noted Jesus. The reward, of course, was merely the public acclaim and subsequent ego-inflation. Interaction between people is public and its own marvelous game, but that between a person and a field is private and hidden, in a secret place—and perhaps a larger, more embracing game in the end. Every negative thought I entertain in my head, which I think is my own secret place, actually strengthens the negative field that sweeps our world, unbeknownst to me—the secret negative thought is shouted from the housetops. Every time I bemoan the negative world out there that I must suffer, I have supported and contributed to it through my moaning. My secret place in my head is not so secret after all.

Similarly, I employ the field of mathematics every time I try to balance my checkbook, and, unbeknownst to me, feed back into that field my clumsy efforts. This may, in fact, be the same generic field formed from the crude finger counting of our ancient ancestors and built up through the ages by Pythagoras, Cantor, Wittgenstein, Poincaré, Einstein, and on and on. Any expression of math might express its field, but no expression fully expresses math or exhausts the field. Every interaction enhances the field from which that action springs or to which it relates. As long as they are employed and translated by neural fields, fields of potential grow—more and more.

Culture itself is a field, an aggregate of ideas, a taxonomy that lifts disparate notions into a coherent and powerful whole. A countercultural movement always strengthens culture though it might tumble some current cultural kingpins. Following such a revolution, culture remains untouched. The field as itself is inviolate, its contents incidental, for it can absorb any content into its own formal elements and subsequently transform it. We thought we would bring true change and affect history with our Vietnam protests, but the cancerous violence of that war simply shifted locale and is more widespread now than ever. Culture perfectly expresses our great model's suggestion to "agree quickly with your adversary"—it automatically makes an ally of every effort to change it and thus has no adversaries. Like the heart, culture can take on any form. Our nation, with its Bibles, flags, and munitions, is but one expression of culture, a field with many faces, using everything to its advantage, growing more and more.

FIELDS OF INTELLIGENCE AND SAVANT SYNDROME

In my last book, *Evolution's End,* I devoted a lengthy opening chapter to what was long called the idiot savant syndrome. I did this to describe the nature and function of fields of intelligence or knowledge, which offers an explanation for the savant phenomenon, a phenomenon that, in turn, gives some insight into the function of fields. Today the term *idiot savant* has been graced with the much less offensive and more all-encompassing *savant syndrome,* in part because the capacity is found in a wide variety of people. My interest has been in the idiot savants as they were originally identified: individuals with an IQ of about 25 who were unable to read or write, were in some cases unable to feed or clothe themselves, and were generally institutionalized in early childhood. An idiot savant, different from the common, garden-variety idiot, though a genuine idiot in most respects, can give volumes of information in one or several subject areas and/or can accomplish incredible mental or physical feats in one particular area.

From careful research we know the savant is truly uneducable and severely disabled and thus can't be taught or trained in the subject he can quote so exhaustively or the skill he can perform so expertly. From the few whose history is known, we find that some event of earliest childhood seems to have triggered or activated a particular neural field that then becomes sensitive to and can translate from a corresponding field or body of knowledge resonant with that event. The savant effect itself is then activated only by someone asking the individual a question related to the available body of knowledge or by the savant entering a situation that is resonant with his special ability. If his neural field is stimulated by sensory input of a similar nature from another person or situation, the savant responds reflexively. Ordinarily the savant can't activate his capacity himself or reflect on it. Essentially his capacity happens to him, or through him.

Further, the field that an idiot savant might access can be quite up to date—for instance, an "automobile savant" who could not feed or dress himself could, from a single glance at a parking lot filled with cars, recount the make, model, and model year of every car there, including the latest editions hot off the assembly lines from factories around the world. Investigation would prove he had no access to such information and couldn't grasp such facts if he had.

Neuroscientist Richard Restak spoke of stimuli "coming in from higher up the evolutionary stream." The savant's source of information is not through his basic sensory-motor system but rather through his high neocortex, a neural network with so few developed fields (thus his idiocy) that no associative thinking can interfere with the flow of information or activity, or bring the paralysis of doubt. Thus the few translations possible are direct and uncluttered.

As an example, a mathematical savant who can't add two plus two on paper or understand the concept of number can, without hesitation, give the answers to complex arithmetical problems posed to him. Some of these answers may run into dozens of digits and may be so involved that they come to the savant hours after the question was put to him by an investigator and can be checked only by the computation of high-powered computers.

For instance, a mathematical savant was shown a checkerboard with its sixty-four squares and was asked how many grains of rice we would have on the final square if we started with a single grain at the top left corner and doubled them on each square. The answer is 2^{64}, or eighteen quintillion and lots of quadrillions, trillions, billions, and so on, which my computer can't fill in. The number looks like this: 18,000,000,000,000,000,000—twenty digits long. It took the savant forty-five seconds to spell out the figure, which he did numeral by numeral.

Such savants are unaware of any computation of their own going on—

the field is an intelligent action responding to stimulus and the savant is simply an innocent bystander in the process. The answer arrives in complete, digital, linear form, and in giving it the savants spell out each digit, using no mathematical language such as exponential "shortcuts." Because they answer number problems only and not algebraic or other symbolic abstract problems (as far as I know), the involved neural fields of such savants need respond only to concrete numbers such as one through ten and thus answers are always presented in this same form, regardless of length.

Savants with other knowledge areas exhibit similar limited yet vast capability. A geographical savant was found who could rattle off the names of hundreds of mountain ranges, the rivers associated with each, their precise longitude and latitude and location in relation to other ranges, and so on, but only if asked to do so, and if the area, country, or continent was designated. He couldn't read or write and had lived in a gray-walled institution all his life. He had no other ability and a working vocabulary of just fifty-nine words.

The famous identical-twin "calendrical savants," George and Clarence, could give any calendar day or date for any event a questioner selected, such as the day of the week over the next several centuries on which a birthday might fall; the exact dates of the seasonal equinoxes for a future year; or the date of Easter in some past or future year (a difficult computation involving, among other variables, phases of the moon). If requested, they could include all information a calendar might offer concerning a particular date, such as the times of sunrise and sunset, phase of the moon, or data on tides. The savants' computations could range some forty thousand years forward or backward in time and would shift to accommodate changes of calendar systems in Europe. They couldn't, however, understand the meaning of *calendrical system*, and weren't aware of shifting their calculations from one system to another. Nor could they do any of this unless given a target date.

The stimulus was environmental, sensory-motor, coming from outside; the response was neural-vocal, from the inside, expressed outwardly. Their sensory system picked up the frequency of the question coming from another person; their neural fields, resonant with that stimuli, responded by tuning into the resonant frequency of the corresponding nonlocal field or frequency field. As recipients of information through this resonance response, savants are probably the most bona fide channelers around.[3]

But how is a savant's area of knowledge determined in the first place? In the case of the twins, there is more to the story. In their impoverished early childhood, George and Charles had an ingenious nineteenth-century novelty to play with, a little brass device of cogged wheels turned by a crank. Each wheel had a set of numbers arranged around its circumference so that when the crank was turned, the wheels and their symbols lined up and any date within a two-hundred-year radius could be determined.

The twins couldn't read the various numbers, of course, and knew nothing of the device's purpose. But on a gray day in a gray life, with nothing else to play with, there was that little machine with its crank. It was a perpetual calendar, a popular novelty in their grandmother's day. In their neural systems that received sparse stimuli and few models, a correspondingly sparse number of neural fields and their structures of knowledge were built up. Repetitive action with the little brass toy brought a neural structure cued to calendars—not a neural field of knowledge of calendars, which would imply content, but a neural field sensitive to a calendrical cue from the environment, such as someone asking a question resonant with that field. Once cued, the field responded in kind through the dynamic afforded.

With few other neural structures functioning and no higher centers bringing discrimination or judgment, with no associative thinking to introduce doubt and the constraints of caution or concern over error, the field effect was wide open. Such radical openness led to the unconditional nature of the twins' knowledge about calendars, a kind of unconflicted neural behavior. None of this explains where the information came from or where the fields are, only that the function exists. A field *exists only as function*, and the function exists only if a question is asked resonant with that field. The question brings the field into existence as the answer. The answer is the field's existence or being, just as my longing for God brings God into existence in my life.

If there is no question asked of a savant or no environmental stimulus is given, however, there is no field. This returns us to a subject brought up

---

3. How could a verbal question trigger a neural field to trigger a corresponding frequency field into dynamic interaction? Sound is macrocosmic and localized, field is microcosmic and nonlocalized. Perhaps coherence is the medium of exchange.

in the beginning of this chapter—our ceaseless effort, however futile, to discover and define the full dimensions of self and God, our longing for closure to what has been an endless, open-ended search. Thus Sat Prem, disciple of Aurobindo and the Mother, said of the spiritual search: "If you are thirsty, the river comes to you. If you are not thirsty, there is no river," and thus the Psalmist cries out that his heart longs for God like the hart (deer) in a chase longs for water, and thus Kierkegaard observes that longing is God's gift.

The twins were institutionalized at about age five. Later their odd capacity was discovered and written up and they were tested time and again. Attention came to them, which they liked, and they eagerly responded to the questions that switched on a mental action they could not switch on themselves, all of which enhanced the field action.

Darrell Treffert, a doctor working with retarded people for thirty years, discovered many idiot savants with all varieties of capabilities. His book, *Extraordinary People*, gives detailed descriptions of dozens of them, but neither Treffert nor anyone else has given a satisfactory explanation for the phenomenon itself—and thus it remains an enigma, one that is not seriously investigated and is avoided by science in general because, perhaps, the premises on which some science is based might be questioned or even threatened by what is found.

THE FIELD AS A UNIVERSAL

▼

A field is a kind of universal for its particular category of phenomenon, and the answers or responses drawn from a field are the specifics or diversity of that universal. A field, like a universal, doesn't exist—it can't set itself apart from itself. Its existence lies in its diverse expression. The problem a savant presents is that the specific field involved in the savant's information area solves problems and even extremely difficult problems that take time to be solved. This indicates that fields of information or knowledge are not just inert repositories of facts, but rather active participants in the dynamic. The field in the savant's case gives calculations only a high-powered computer could duplicate, and can even express information available as recently as the present moment of the savant's response, information to which he has no physical connection.

Esoteric philosophy speaks of the Akashic Record—supposedly a complete record of our species' past—from which we can draw. We could attribute such a record to fields of memory, but memory is of past events and the savant function shows both memory and intelligent action taking place in the moment, with current, even future events at its disposal. So this notion about the Akashic Record clearly indicates field effect, but can't explain it—all of which borders on the idea of *mind at large* found so often in the perennial philosophy.

NO MAN HAS SEEN A FIELD

▼

Because a universal or field can't exist, or be set apart from itself, we see the accuracy of the biblical adage that no man has ever seen God and lived. Nothing in existence can "see God," who doesn't exist. We can, however, see parts set apart from that universal and thus see God in each other, as Muktananda challenged us to do and as Jesus suggested when he said "no man has seen the Father" and then pointed out that "he who sees me sees the Father"— which seems about as contradictory and paradoxical as possible.

Michael Sells calls such verbal contradictions *unsayings,* the classical language of the mystics. Unsayings are logical reversals within the same statement that attempt to describe the indescribable and the seemingly illogical. Unsayings attempt to take us out of our ordinary mind-set and open us up to who we are beyond the cultural self-definition we've formed from our birth. How exactly do you express the inexpressible or define the infinite? The unsaying is the closest mystics can come to putting into language that which can't be spoken. Words are available to refer to anything that exists, but the mystic points to the other side of the coin. The Tao can't be spoken, said Lao-tzu, and then he proceeded, with some sixty-five verses, to speak about the Tao and explain why words had nothing to do with it.

As with most forces, we know fields not as themselves alone but through the results of their function or actions. No man has seen gravity, but we assume there is a force we call gravity because things fall to center. Rather than setting out some eightfold abstract prescription for behavior or an outline of God's plan for salvation etched on stone tablets, Jesus simply said he was the way—he was It, or, as Eckhart would contend, its mode of being. Jesus wasn't talking about It, which can't be talked about—he *was* It, or rather, It was he, and those around him could either see that or not. A field flowed through him and moved him, but he was not the field. In seeing him, people saw the field in the only way fields can be seen.

Jesus modeled the function he pointed toward, which is all that can be done concerning field function. His way was a function that exists only as we initiate and activate it. Like manna from heaven, it is given in the moment and can't be stored up, suggesting that any form of institution concerning his way isn't that way. We can understand his way only by becoming it, by letting ourselves, allowing ourselves to be an expression of the field.

The spirit of wholeness, the Paraclete (helper, guiding spirit) Jesus bequeathed us, is its own field and field effect, a combination of intelligence and power. It flows through us and possesses us, but is not ours to possess.

In the same way, in Eastern spiritual systems the energy, or Shakti, of the system flows through an individual but can never be a personal possession. It travels along its own highly selective channels through the gurus (the word, like *rabbi*, means "teacher"), who are the few people who fully allow it to flow. When Baba Muktananda put his hand on my head at my crown chakra, a neural storm could break and often did. When it did, I was not likely to forget it, for my perception was changed each time. This Shakti or power of spirit never manifested twice in the same way and always caught me by surprise. To this day I find myself trying to get a handle on that Shakti and make it mine. But I can't possess it—I can only be possessed by it. Trying to make it mine kills the dynamic and I am left with dust and ashes instead of a river flowing through.

Jesus' Shakti, or Holy Spirit, should have been built up in this way, an individual's personal charge attracting his far greater one, resulting in both being enhanced. But instead an institution was set up to possess, dispense, and profit from his Shakti—and the Shakti was no longer there.

All human experience, whether mental, physical, or emotional, will, if repeated, aggregate in field effects. These can then act in a causal way, leaving their impression on our minds, which we, in turn, act out. William James commented that our reality is merely a minor selection from a vast potential that exists all around us and is separated from us by the "flimsiest of veils"—an observation resulting from a drug experience during which that veil was rent. The only value in such a rending of the veil would be in letting us know of that vast, unboundaried potential all around us. This potential would overwhelm us were not our brain carefully designed by nature selectively to screen out fields that have no relevance and open to those that have. Schizophrenics may be those whose veil has been permanently rent, while idiot savants have particularized holes in their veil over which they have no jurisdiction. Rudolf Steiner, like Jesus and other "genius savants," could selectively open his veil and access that infinite realm in a highly selective, intelligent way, which is no doubt where our general evolution should lead.

BIRD BRAINS AND FIELDS OF KNOWLEDGE

▼

Neuroscientists Eric Knudsen, Sascha deLac, and Stephen Esterly, considering the computational maps of the brain, stated that no matter what our perceptual experience, no matter which senses are involved, any experienced sensation we have can be represented as a "peak of activity from a population of neurons"—experience is a result of a neural process within us. If there is no neural activity, there is no experience of body or world.

In a recent (1997) paper, neuroscientist Paul MacLean, going strong in his eighth decade, makes a similar observation: ". . . we can never discover anything outside the brain because all the ingredients of cerebration, like those of a mathematician's formulas, are already in the brain." The ingredients for cerebration are in the brain, while that which they translate is both within it and outside of it. The tiniest change in neural current can bring a dramatic change in what we are experiencing "out there."

Our body and brain form an intricate web of coherent frequencies organized to translate other frequencies and nestled within a nested hierarchy of universal frequencies, all functioning in coherent resonances. MacLean quotes Vandervert from his *Neurological Positivism*, referring to "science's erroneous placement of the world outside the skull." All our experience is due, MacLean states, to "what we realize by means of built-in algorithms of the brain." To which I would add that the algorithms are built in and the materials on which they operate unfold moment by moment through rich dynamics between possibility and actualization.

Only the actualization, the final objectified experience or world, can be called "outside the skull," although it obviously isn't outside any more than inside. The evolution of our brain has led us to our ability to create this magnificent universe as an event outside the skull, which we can thus perceive as an object even though it remains an event that we can never prove or disprove as either "out there" or a projection from "in here." In this way our reality proceeds as an open-ended dialogue between creator and created.

When we look out at the farthest nebulae of stars in some remote corner of the universe untold light-years away according to our conceptual interpretation, one complex of wave fields is interpreting another. Ornithologists have speculated on a bird's capacity for long-distance sensing—how, as one proposed, a bird flying over Ithaca, New York, for instance, seems aware of affairs on the coastline a couple of hundred miles away. Consider that the bird's neural system selects from a universal pool of frequencies those specific to that breed of bird's well-being, precisely as our brains do, or were designed to do. There are, then, quite deliberate designs rather than chance selection to guide body-brain systems. A bird's world is made of selections that might have only the slightest relation to ours.

When a hawk spots a mouse from a great distance, we marvel, but distance is relative and the hawk's brain highly selective. A hawk might not be seeing the bumblebee two inches from the mouse, nor the daisy close nearby, at least not with the same clarity with which he sees the mouse. These peripheral sights would clutter the scene and obscure the important mousy things of life. The hawk's glia select those frequencies his neurons need for his world experience and well-being in it. Biologists Humberto Maturana and Francesco Varela described how the eyes see what the brain is doing—even as the brain is doing according to what the eyes see. This is yet another expression of our reciprocal dynamic.

Consider the medical account of a man dying of a brain tumor who, in his last few days of life, displayed remarkable vision. He could actually see the fleas on a dog blocks away, and created quite a stir with this remarkable ability. His visual system functioned as usual, but selectivity and the orderly feedback maintaining perspective and similar perceptual details had broken down. He was seeing the way a hawk does, perhaps.

In her last days, my aged and dying sister grew excited over various gnomes, elves, and other creatures in the tree outside her window. Her eyes were seeing what her brain was doing and vice versa, but her brain was

doing extraordinary things, pulling stuff out of childhood memory perhaps, with no concern over appropriateness or logic. Her visual selectivity
had slipped a cog—or so it seemed to me, who, in my blindness, could not
see Peter Pan in the high limbs of that oak tree.

THE VIRGIN WHO SWINGS THE SUN

▼

At Medjagorge, in Yugoslavia, the Marian phenomenon (appearance of
the Virgin Mary) has been occurring since the late 1970s. Fifty thousand
faithful and curious from all over the world crowd together each day to
experience a presence found only there and to watch the capricious behaviors of the sun that can be seen only from there.

A most skeptical BBC television cameraman, assigned to film the miraculous, turned to his colleague and boss to report that he had stared un-
blinking at a hot Mediterranean afternoon sun for twenty minutes and had
no afterimage blind spot nor impaired eyesight. Shortly afterward, all fifty
thousand mind-brain-bodies present witnessed the sun begin its astound-
ing swing in slow arcs over the sky, a feat it has been performing in that
locale the past two decades.

In response to the sun's sweeping arcs, the same cameraman turned
again and asked of his colleague, tentatively, "Could you possibly be seeing
what I am seeing?" Indeed, all present were. Only in that one locale does
the phenomenon take place, however, perhaps as a kind of substitute for
the appearance of the Virgin herself, years after her first visitation to young
goatherds there. Cameras, mechanical devices, can't "see" in this way; they
have no neural systems subject to variation, so they can't enter into the
dynamics involved. A camera "sees" only stimuli from the earth as itself. It
has no brain and heart to open it to higher realms of vision. Scientism
claims that only the camera sees the real world, while fifty thousand seeing
a miracle unfold are deluded. Blake urged us to use our eyes to see *with* in
creative vision, rather than simply *through*, as we look through a pane of glass.

Consider the brain's micro-millivolts of neural interaction mentioned
in chapter 3. It would take but the minutest bit of energy to shift the infrared and visible light frequencies of that sun a tiny fraction in a particular
locus. All that's needed is a tiny shift to bring about a dramatic alteration of
ordinary visual process in one or any number of participants simultaneously.

There is only one heart, as my meditation teacher Gurumayi reminded us. Each brain translates its reality according to the locus environment and participates holographically with that spectrum of sun—and events move accordingly. But nothing changes within either the sun or the universal heart; the only change occurs in our highly malleable, adaptable, flexible, and creative brains that have evolved for just such infinite games.

Again, in the electromagnetic spectrum, voltage power is not an issue in matters of conscious awareness and perception, but translation of frequencies into experience is. To ask "Does the sun really swing?" in Medjagorge is to miss the greater issue of dynamic systems. In Medjagorge the eyes see what the brain is doing and the brain pursues its doing according to what the eyes see, a dynamic in which each brings the other into being. There is no being except in a mode of being. William Blake insisted that God was because we are. And Eckhart would agree that without us, God is not, but would add, with Blake, that this doesn't mean we are God.

A radio receiver is a critical part of the radio world—without a receiver the rest of a broadcasting system is worthless. The receiver alone gives those invisible waves their being throughout that broadcast field. But the receiver is not the sending station nor the field itself. The same is as true in the labyrinths of our brain-mind as in the world of radio. A field effect might be registered by a particular brain-body receiver, producing a corresponding experience, but that brain-body, while one of an infinite number of possible loci, is not then *the* locus, genesis, or possessor of that field. Bernadette Roberts points out that our experience of God, though of shattering, life-changing dimensions, is one of our experiences, not God's.

Eckhart said that as he became God, God became Eckhart—rather as though they gave each other being. But in the final analysis Eckhart could make no statement about God as God, as I can make none about mouse as mouse. He could only speak about God "insofar as God is Eckhart" and Eckhart "insofar as he is God," to use Eckhart's language. Following Eckhart far enough leads, however, to that disturbing chasm of unknowing he points to in his prayer "Oh God, deliver me from God!" The God Eckhart prays to be delivered from is God-as-Eckhart. God-as-God can't be named, cognized, or described as can God-as-Eckhart, which leaves only God-as-God aware of God-as-God.

Plotinus spoke of a "superplenitude of love" creating universes to express that love, which love always exceeds its ability to be expressed. (It is thus, I assume, that the universe expands.) Yet that universe can experience only the feedback of love within those universes of dynamics, because such is all that love can know, so to speak—it's a matter of resonance. Just as the camera couldn't register the swinging of the Virgin's sun in Medjagorge, love can't know what is not its own. Love is its own field, in effect, which can be brought into being only through its own wavelength—it can neither be nor be aware of something opposite to itself and so "judgeth no man." Love has no being except through a mode of being, yet can't resonate with anything but its own nature reflected by its reflective system for being, just as I can't know the experience of a bird that wings the airy way.

Thus Suzanne Segal, having fused with the universal, experienced that universe as itself, and claimed that the "vastness doesn't know anything is wrong." This resonates with Jesus' claim that his Father judged no one and "rained on just and unjust alike." If understood and accepted, these statements would eliminate religion and culture from the human experience, leaving us open to spirit and truth. Jesus brought to our awareness a God of love who gave only "good and perfect gifts." But a frequency can only perceive a corresponding like frequency, and for us to perceive and interact in dynamic with such a God we must function as that frequency. To do so, as Jesus both stated and modeled, is to "do the Father's will."

Mae Wan Ho found a dynamic, responsive, liquid crystalline medium that pervades the whole. This power of wholeness or holy spirit can take on any form and is always there within our experience, whether in a laboratory or on the farthest star. Eckhart, having experienced that pure realm beyond all diversity, prays to be taken up into it again. God-as-Eckhart was a lofty state, but apparently nothing compared with God-as-God. Segal fused with her universal, and perceiving It perceiving within her exclaims over and over that we are not the doers—It does everything. But only as It did everything through Segal was she aware of that universal. "Not I but the Father within me does these things" said Jesus of his miracles. But without Jesus, there is no Father and there are no miracles. "Without me, God is helpless," was Eckhart's audacious claim, in reflective resonance with Suzanne Segal, and the everlasting gospel.

PART TWO

▼

THE ANATOMY OF EVIL

PREFACE TO PART TWO

▼

DORCHESTER HILL

In retrospect, something akin to spiritual awareness seems to have arisen in my fifth year in conjunction with two specific events. The first was daybreak. The first faint light of morning brought to me waves of some ancient knowing that produced a lump in my throat, a kind of homesickness, a shadowy remembrance—of what, I did not know. I only knew that the rising sun stirred an intense longing in me, which I usually was up and about to experience (and still am). By my seventh year I always tried to be on Dorchester Hill at daybreak, weather permitting. This windswept and barren knoll rose not far beyond our house, and from its pinnacle you could see the sweep of the whole valley and the surrounding Appalachians.

As I approached this high vantage point, the longing within me intensified. Making my way up the last yards of the slope, I had the clear impression of an invisible veil, like a glass wall, separating me from it. Nearing the summit my heart would pound, for *this* time I would break through and the object of my memory and knowing would be revealed—a thought that held a peculiar dread for me as well as exercising a compulsive pull.

But each morning I crested the hill and began my descent uneventfully, and disappointment would wash over me anew—again the veil had not been rent. Nothing had been revealed, and it never was.

The second event that marked some form of spiritual awareness of my childhood was a recurrent night terror, which also began when I was five years old and persisted until I was eleven. A night terror differs from a nightmare in that once it begins, it will continue until it runs its course, even when the dreamer's eyes are open and he seems to be awake. For someone experiencing a night terror, visual surroundings are taken in through the eyes and are absorbed by the inner vision, becoming part of it. Any external sensory stimulus is then able to reinforce the dreamer's inner imaginary state. As a result, a person can't be awakened from a night terror; any such attempts made by an outsider often become part of the context of the "dreamer's" inner experience.

My dream began with a muffled, pulsing drumbeat, both heard and felt, that grew in intensity until it filled the whole universe, absorbing everything and becoming inconceivably immense and totally terrifying. At that point a peculiar, metallic, nonhuman voice began questioning me steadily, repetitively, its demands growing louder and louder until the voice and the great immensity became one pulsing sensation. (In early childhood I was terrified by any loud sound—and the loudness of this voice was beyond all conception.) The question itself was simple: "What is it? What is it?" asked over and over again. Any exterior sounds, such as the talking of my family around me, and my own screaming (which began immediately), became part of that pulsing immensity, magnifying it. This event always ran its course, the sounds eventually diminishing and finally leaving me. My outer world—including my whole family clustered around me, trying to stop the screaming—began to reassemble, and I was left feeling exhausted, nauseated, and shaken for days.

As is usually the case for night terrors, and as I was told many times afterward, my eyes were open throughout the episode, though what I saw was not my family but an image that was quite difficult to describe because it was equal parts auditory, visual, and conceptual, all centered on the repeated question, "What is it?" The great immensity in my dream was a pulsing red amorphous thing filling space with its integral parts of sound, substance, and repeated demanding question.

The dream was followed by a peculiar amnesia that wiped out all recollection of what the episode had been, though it would haunt me for days afterward, making me dreadfully afraid of being alone. It recurred three or four times a year until I was eleven, with its impact worsening as I grew older until it became a principal issue in my life. When I was eleven, however, it occurred to me that I could, with effort, ride through the dream without losing myself to it; I could stand outside it, remaining conscious, and in this way remember its contents, for I was convinced that if I could just remember what had happened, I would be free of it. I kept paper and pencil next to my bed and each night would reaffirm my determination to remember the dream if it occurred again. According to Jean Piaget, it is at or near age eleven that formal operational thinking becomes possible. To use his terms, this stage of growth is characterized by the capacity of the mind to stand outside brain function and "operate" on it. And this is precisely what I did when the dream finally recurred—indeed, I found myself subjected to it and absorbed by it, and yet, on some level, remaining outside of it where I could observe it. It was this outside observer in me that prevented me from screaming, which my whole body tried to do, while I, aware of each tiny piece in the scene, rode the experience through to its end. After it had passed I immediately wrote down a full description of the event and learned that I could remember it with remarkable detail. I never experienced it again after that, and on a couple of occasions when I felt something similar to that state approaching, I went along with it, at which point it evaporated or dissipated.

Throughout my childhood I was a passionate lover of the Episcopal Church and the hottest acolyte in the Southwest Virginia Diocese. As in the Roman Catholic Church, the acolyte assists the priest in the various ceremonials at the altar, which, altogether, makes for stately, dignified, and dramatic pageantry. At age twelve the diocese offered me a scholarship to the best prep school in the South, which would be followed by the university and, finally, the seminary so that I might be frocked and shepherd a flock. I was elated but my mother was scornful. She refused the offer for me, reminding me that ours was a family of honorable newspaper people, that her brothers and father, like my father, had been country weekly editors, city editors, writers. The same country weekly had been in her family for nearly a century. With the greatest scorn she said: "Joseph, all nice people go to church, but they don't let it go to their heads." And I was not to let it go to mine.

In mid-adolescence, three states or conditions of mind were central to my life. The first was an intense idealism, a noble set of standards so lofty only Jesus could have or did live up to them. (Surely not I nor anyone I knew could.) A second was my affliction with what I later termed *hidden greatness,* a constant, exuberant bubbling up of my own enormous, exultant personal stature and importance, a depth and magnificence of being within me that no one out there in the world could possibly detect—I was then, as always, quite small in size and profoundly nondescript in appearance. I avidly adopted Walt Whitman's lines: "Dazzling and tremendous, how quick the sunrise would kill me, / If I could not now and always send sunrise out of me." This sunrise manifested as an irrepressible exultancy deep within my being and burst out of me like a shout.

The third state that defined my adolescence was even more intense. I later labeled this *the Great Expectation.* An ever-present conviction that something tremendous was supposed to happen existed within me and continually grew stronger. And it—whatever it was—was supposed to happen immediately, this very day, this hour, this second; it was right around the corner, over the next hill. Decades later, following a talk in Santa Fe, New Mexico, a couple shared with me a letter from their son, a junior in an eastern university, a top scholar, athlete, big man on campus. His letter addressed, he said, an issue of such magnitude he could trust sharing it only with his parents, and I felt honored they had shared the letter with me.

He had awakened in the middle of the night, he wrote, with "the cold hand of terror" clutching his heart. Ever since he was about fourteen, he continued, he had been waiting for something tremendous that he knew was supposed to happen. It was something so huge and ultimately important that there was no way he could speak of it and instead had quietly nursed the longing in his heart. The terror seizing him in the night, he explained, was his approaching twenty-first birthday and his realization that for seven years he had been waiting for *it* to happen—but it had not happened, and in that dark moment he knew that it was never going to happen. "I can live with the fact that it will never happen," he concluded, "but find it difficult to accept that I shall never even know what it was supposed to be." Long before my own twenty-first year, the great expectation within me was usurped by World War II and the Army Air Corps, where I spent my late adolescence. There I sat, through many a "Why-We-Fight" film, complete with the atrocities designed to incite us future pilots and bombardiers toward the mass murder required of all good airmen. Finally, when the worst of the most intense atrocities were displayed on that screen they were accompanied by Sergey Rachmaninoff's poignant, emotional orchestral work "The Isle of the Dead," and I wept openly. One of my cadet-trainee buddies seated next to me hissed in my ear to shut up—if the attending sergeant heard me, I would be immediately washed out of flight training.

During these months, in rare spare moments, I read snatches of Will Durant's view of history, and in light of Will and and his wife Ariel's wisdom and the horrors viewed on screen, I dutifully became an atheist. Secretly, however, I held to a love of Jesus and a long-cherished romantic image of him, a kind of closet affair of the heart that had grown over the years. Of God I had my severe doubts; of Jesus as the greatest of humans and model for us all, I had none.

▼

FIVE

WHY NATURE'S PLAN BREAKS DOWN

The Absolute, to man, is his own nature.

—LUDWIG FEUERBACH

One of Paul MacLean's most valuable contributions was his insight into what he termed *the family triad of needs:* audiovisual communication, nurturing, and play. As with all mammals, our human nature rests on these three interdependent requirements, without which we could not long survive as a species. These needs bring about and sustain human development from birth and are, I would add, the springboard to transcendence itself. Our failure to provide all three disrupts intelligence and social development but at the same time supplies the means for enculturating us, thereby sustaining culture.

In brief we can say audiovisual communication is required by the R-system in its connections with the emotional system. This need is met through the bonding of infant and mother. Our critical need for nurturing and the pleasure we take from it are built into the emotional-cognitive system and its links with the heart and prefrontal lobes. This need, too, is met through the bonding of infant and mother. When the requirements for audiovisual communication are met, they automatically fill the needs for nurturing—communication and nurturing spring from the same interaction, and, in effect, bring each other into being at birth.

This rich dynamic gives rise to play, the need of our creative-intellectual neocortex and its connections with our emotional system. Play unfolds in the safe space created by audiovisual communication and nurturing—the safe space that results when all our needs are met; the safe space that both brings about and is brought about by the bonding of heart and mind, which, in turn, results from the bonding of infant and mother. In this safe space where censure cannot occur because error doesn't exist and where time is not a factor, play can be freely established.

THE MODEL IMPERATIVE

▼

The family triad includes by default nature's imperative that a model be given for all aspects of development. Recall that a model is the living embodiment of the child's inherited capacity or talent and that its stimulus—a possibility demonstrated by the model's presence—brings about a like response in the child, building a structure of knowledge, or imprint, within him.

There are no exceptions to this necessity for modeling, and three examples are presented here: our capacities for language and vision and intelligence of the heart. These three unfold as naturally as breathing, are neural imprints, or constructions of knowledge we automatically make, and their need for ongoing model stimuli is exemplary of all of our capacities.

Back in the 1940s Bernard and Sontag published research on fetal movements made in response to sound that were detectable from about the fifth month in utero. Subsequently, in the early 1970s Boston University's William Condon and Lewis Sander discovered that at birth the newborn responds with a precise muscle or muscle group to each of the phonemes used in the mother's speech. (A phoneme is the smallest part of speech making up words. Our alphabet is phonetic.) From a newborn's repertoire of movements, Condon and Sander were able to map out all of these muscular responses, which form in utero and are well established by birth. They could analyze this synchrony between phoneme and movement and accurately predict the infant's muscular response to any word spoken in the infant's immediate presence.

The stimulus of this dynamic is the mother's speech, the response is the infant's muscular movement. Because this imprint begins in the late part of the second trimester of gestation and is largely established by birth, the capacity is clearly "hardwired" as a genetic potential. Genetic hardwiring is only half the story, however. If there is no speech stimulus from the mother, there is no development of phonetic response in utero, no matter how powerful or selfish the genes for this might be. If the infant is deaf or the mother is a deaf-mute, the infant is born without the muscle-phoneme synchrony.

A hearing infant of a mute mother must eventually be in a language environment long enough to first build those muscular responses; the rest will follow. In the same way, every perceptual and conceptual structure of the brain forms in response to a like stimulus from a model in our environment. The first imperative of nature is simple as rain, and as natural: no model, no development.[1]

Alfred Tomatis, a French physician studying the effects of sound and speech on the nervous system, made the same discoveries regarding the phoneme-movement synchrony at about the same time as Condon and Sander. Tomatis's research showed how every cell of the skin is an "ear" that picks up sound waves. Other research revealed that each stratum of muscle fiber in the body has a muscle spindle on it, a minuscule nerve ending connected to the peripheral nervous system and to the cerebellum in our head. (The cerebellum is the brain module through which muscular movement, including that of the many muscles involved in speech, is initiated and coordinated.[2])

This sound-sensitive network of listening cells, muscle spindles, and cerebellum in turn becomes selectively sensitive to speech sounds, and then to single phonemes. All development seems to move from the broad and generic to the more singular and specific. The same phoneme pool, or field, consisting of forty-two units, underlies all language. Every culture draws

---

1. Overlooked in the story of Helen Keller is that Helen was not born blind and deaf, and thus her teacher was not starting with a "blank state." Helen was an ordinary bright, hearing and speaking child until her eighteenth month of life (nearing the end of her toddler period) when she developed scarlet fever and lost her senses of sight and hearing. By the eighteenth month close to 50 percent of a child's full sensory world, particularly language, has been roughed in and, to an unknown extent, myelinated. While Helen's nurturing and reawakening is a wonderful example of the model imperative, bear in mind that all those structures of knowledge that developed in her first eighteen months, including all the neural wiring involved, would have been largely intact, with only the connections to the outer world broken. An infant born without such connections is at a far greater disadvantage, and the child's compensation for this would follow a more difficult pattern than Helen's.

2. Tomatis found certain extremely low frequencies that can confuse the brain as to whether the signal is something to be heard (indicating sound) or felt (indicating matter). The eighth-century cosmology called Kashmir Shaivism claimed that sound came first in creation, followed by light, then matter. Physicist David Bohm called matter frozen light.

on this same pool, or generic field, according to the individual speech pattern in its language. (Some cultures use as few as sixteen phonemes, some use all.) This unique phonetic aggregate is then both the universal and the specific pool for each infant born into a certain culture, and elicits from the infant his unique repertoire of muscular responses. This clearly depicts how unity gives rise to diversity, or how the universal becomes specific in endless variation.

Marshall Klaus, an enlightened obstetrician at Case Western Reserve Hospital, made movies of newborns in which the infant repeats the same, unique movements over and over, as long as the same words are repeated. (I had heard of Klaus's marvelous films for years but never actually saw them until 1998, at a birth conference in Chiang Mai, Thailand, where both Klaus and I presented.) Through speech, then, Klaus could initiate a fine little infant "ballet." Although these movements, easily observed after birth, rapidly become microkinetic, too small to be visible, they are detectable by instruments throughout life.

All learning and growth follow this pattern of unique phoneme aggregates from the same pool eliciting unique muscular movements from an array of possible muscular responses. Our growth moves from universals to ever more precise and specific individual variants as a result of following environmental models. William Blake said the broad and general, or universal, was useless. Only the particular, specific, and concrete is meaningful. Thus a wispy, ethereal, universal god becomes meaningful to us only as made real and specific, one among us.

Not surprisingly, if born to or raised by a French-speaking mother, a child will speak French, while a Japanese-speaking mother produces a Japanese-speaking child. This obvious truism leads to the second component of nature's model imperative: The character, nature, and quality of any intelligence or ability are determined to an indeterminable extent by the character, nature, and quality of the model. This assumes that the model will not only awaken but also guide the intelligence or ability as it develops. No one-for-one mechanical mirroring is implied because the procedure is stochastic, with an element of chance in its function as in all development, but dependence on stimuli from a model is indisputable.

Our second example of the model imperative at work is our acquisition of vision. At birth the newborn doesn't open its eyes and gaze upon its glorious new world, for no such world yet exists. The infant must first build a visual structure of knowledge of his world, a huge project involving more of the brain than any other activity. While reaching a first level of stability at about nine months after birth, a fully developed visual system takes some twelve years to complete.

To initiate this enormous undertaking, the newborn brain is hardwired to see only one object at birth: a human face, the template on which all vision is then built. Newborns react negatively to bright light and not at all to objects unless the object has a facelike quality or is sufficiently complex to contain the rudiments of facial characteristics. If an infant's visual system is to be stabilized, the face must be presented at a distance of six to twelve inches and remain there for a majority of his waking time. Nature provides a whole cascade of instinctual interactions between infant and mother to ensure this close proximity of a face from the moment of birth.

At birth any face will do, even a false face (for a short time), if presented at the critical distance of six to twelve inches away from the infant's eyes. This literally turns on the visual brain and, as important, awakens general awareness in the infant. Awareness and development begin within minutes after birth—if that opportunity for face imprinting is provided. The newborn will soon display parallax of the eye muscles (coordination between the two eyes) and within minutes will be able to follow a face if that face moves about. Shortly thereafter, as a result of this awakening of the infant's awareness, he will smile at each presentation of that magical face stimulus. Without it the infant slips back into limbo. The infant's responsive smile to the familiar face is also built in, as is the "face's" response to that smile. That infant smile turns us on and locks us into the bond being established.

This period of face imprinting lasts for the first few weeks—the "in-arms" period—and the infant's need for it diminishes on a graded scale throughout the first critical year after birth as the visual system develops. Visual learning unfolds through the child's association of a new, unknown object with the known pattern of face, and goes through many stages that expand visual skill until finally the initial stimulus of a face is no longer needed.

One of the major reasons for an in-arms period in human infancy is to keep a face at the critical distance from the infant's eyes so that the brain is kept awake, turned on, looking and learning. Denied this fundamental opportunity, the infant maintains only a few basic survival reflexes, such as sucking and grasping, and even then he must be given something to suck and grasp.

Beyond its necessary role in the development of vision, another imperative met by the in-arms period and the crucial six-to-twelve-inch distance of the adult face from the infant's is that it keeps the infant's heart within the nearest radius of the mother's heart, as it was in utero. This is critical for development of the primary functions and intelligence of the infant heart in those early months. While in utero, the infant's heart responds to the electrical, hormonal, neural, and sound patterns of the mother's heart, which stimulate and stabilize the infant's heart on all its levels. Completion of this basic heart stabilization follows birth and for some nine months requires frequent periods of close proximity of the infant's heart to the mother's. Recall our example of the two heart cells on the microscope's slide, and the fact that heart-brain entrainment between mother and infant takes place only when they are in close proximity.

All these complex needs of heart, brain, and physical body were worked out by nature long ago, as Nikos Tinbergen's research showed. The results of this research can be summarized as follows: First, nature withholds the production of hydrochloric acid in the human infant's digestive tract for the critical first nine months after birth—the average time both the heart and visual systems need to achieve their first level of stabilization. Hydrochloric acid is necessary to digest fats and proteins. Second, to ensure this first level of development of heart, visual, and sensory systems, nature arranged that human mother's milk be the weakest and most watery of all mammalian milk, virtually fat- and protein-free, though a rich cocktail of hormones. Through this simple omission of hydrochloric acid from the infant's digestive system and the elimination of the need for it through the absence of those ingredients in mother's milk that make it a necessity for digestion, nature arranged that the human infant's metabolic system require nursing about every twenty minutes. Constant mother-infant interaction is thus ensured—and at just the distance to ensure the activation of both visual and heart systems. Sensory stimulus of the new nervous system is likewise ensured by the constant touching involved in this continual interaction.

When the infant is ready to stand up and join the mammalian world, heart, visual, and sensory systems largely functional, nature turns on the hydrochloric acid. Now the newly foraging toddler can digest any proteins and fats he might run across as he charges about, exploring his world. Brilliant planning.

The need to feed their infants so often was solved ages ago by mothers "wearing" their babies, with a simple sling holding baby to breast. Mother could thus quickly return to her normal routines. And indeed, mothers who have given birth free of interventions, crippling drugs, and the use of instruments during delivery are back in circulation within minutes after their babies are born, ready and able to bond with their infants through the constant close contact that provides the newborn the universal safe space and continually varying environment for visual stimuli and construction of his world.

Through the simple, natural, and uncomplicated activity of breast-feeding, nature arranged this failsafe way to take care of the foundation of the family triad of needs. And as the work of a panoply of research people from around the world showed years ago, nature built into the mother an overwhelming, compelling drive to provide for her infant by building into both mother and infant a cascade of powerful instincts to respond appropriately to each other—if given the chance.

This dynamic interaction or bonding between mother and infant is made possible by nature's foresight in equipping mothers with exceptional mammary glands, not primarily to drive men mad, but rather to make bonding even more worthwhile to the mother and easier for the infant. Nature designed nursing to be a gratifying sensual-sexual experience for the mother, as well as an obviously satisfying sensual event for the infant—a real mutual back-scratch. While infants can't articulate reports of their experience, one of the reasons many mothers breast-feed their children long term may be that sexual stimulus and even orgasm from nursing are possible. I suspect women have kept quiet about this added incentive lest nursing be outlawed by some religious groups.

To sum up, then, the importance of the model imperative in an infant's initial development, the mother's voice is the model stimulus in utero, which activates the infant's language and sensory-motor system. In the same way, presentation of the mother's face at birth acts as a stimulus to which the infant responds with awareness and the initial development of vision. (In the case of congenital blindness, nature compensates, as always, as best she can.) And so it goes with all forms of human capacity, whether sensory-motor, emotional-cognitive, or intellectual. Outer stimuli bring inner neural-muscular responses and eventual growth of a structure of knowledge or learning. And in all development, given the appropriate model environment, functions unfold automatically, as nature designed. Denied the model, nature must compensate and the function is compromised.

UNDERMINING THE FAMILY TRIAD OF NEEDS

▼

All systems are dynamics, intricately interdependent and interactive. Environment and genetics are a paired dynamic. As with the creator and the created, they give rise to each other and are interdependent to an indeterminable extent. The mother is the only environment the infant has in utero, the principal environment of the infant for the first nine to twelve months after birth, and a critical part of the child's environment for the first three to four years.

The one aspect of humans that nature couldn't anticipate or prepare for was the development of a male intellect that encroached upon and finally threw monkey wrenches into every aspect of this wonderfully designed birth-and-bonding procedure. This encroachment was slow, devious, and deceptive, but thorough. During the Middle Ages and the emergence of the Inquisition, a growing fanaticism concerning witchcraft centered on the crone, the elderly midwife who passed on to the young women whose infants she delivered the general background of female wisdom handed down through the ages. The crone became a major target of the Inquisition and her body of knowledge suspect.

Among many issues that rankled the cloth was the crone's notion that childbirth was neither a painful nor a dangerous ordeal (and indeed, under the crone's skillful hands it seldom was). After all, churchmen reasoned, the Bible itself said that pain and suffering in childbirth was a sentence pronounced on womanhood by God—of course the crones had to go. Thus their demonization as preparation for their complete elimination became doctrine, and even today the term *crone* brings to mind the archetypal witch, a toothless hag hunched over a fire, stirring a pot of evil brew. As the crone was slowly exterminated, a subject we will touch on briefly again in chapter 8, the surgeons of the time—who pulled teeth, cut hair, and performed various unsavory tasks such as, eventually, dissecting cadavers with the same unwashed tools they used to assist their bungling during births—began to invade the birthing field, supported in their efforts by the cultural powers that be.

From the late Middle Ages on, as detailed by Suzanne Arms in her remarkable book, *Immaculate Deception,* medicine men in general horned in on this mother-child bonding domain. Following Bacon's proposals, dominating nature in all her roles had become the scientific passion (oddly fitting the church doctrine from which the notion arose). After centuries, the practice culminated in modern times in which doctors in twentieth-century America eliminated some 97 percent of breast-feeding and thus the central function around which the multifaceted bonding procedure unfolded. Bonding became the butt of jokes in academic and sophisticated circles and was viewed as a notion adhered to only by hippies and New Agers.

The same pattern has followed in the various countries that have bought into the American way of birth. Thailand and Japan, for instance, adopted our medical practices following our use of these countries as bases for military operations in the middle of the last century. In the spring of 2001 I was asked to return to Thailand for a lecture tour on the issue of the effects of birth and bonding on education—Thailand still has an 80 percent C-section rate, even among its peasantry and mountain people. But I saw the futility of my efforts and declined. (In addition, twenty hours in the air each way is now a bit much for my bundle of bones to bear.) The World Health Organization had sponsored a three-day conference on birth and bonding in Thailand in 1998, which I attended and addressed twice. Very few Thai doctors attended the conference, however, and those few did not stay long. No change in birthing practice has since taken place there, and the family, educational, and social systems of the country have continued their rapid decline—oddly paralleling the same decline in our own country and Japan. I have addressed all of these issues at length in previous books.

In America the disruption of bonding through the elimination of breast-feeding and the separation of mothers from infants during the long hospital stays that were often required through much of the last half of the twentieth century set the stage for Madison Avenue to turn the breast into the hottest sales gimmick ever discovered, an unconscious cultural collusion between two destructive forces: medicine men and advertising men. If denied the breast at birth and during infancy, a male can become obsessed with breasts. Assuming that marriage assures him permanent rights to a pair, he can become unhinged when an infant comes on the scene and takes over, particularly if the mother breast-feeds. Some fathers object to breast-feeding, which is hardly supportive for the mother or surprising for males who were not breast-fed and nurtured themselves. And some of these men, feeling abandoned yet again, may in turn abandon their families—another cultural double bind wherein everyone loses.

Since separation of mother and infant became *de rigueur* practice throughout most of the twentieth century, many of our infants, to say the least, were not given appropriate nurturing or provided appropriate models and stimuli at birth or in the critical first year. For decades, then, a newborn's vision was likely to be restricted to masked faces in those critical first hours, their movement and world restricted to bassinets and cribs throughout the early months, nursing reduced to the solitary experience of bottle-feeding, all too often through a bottle holder. The result was that new infants remained essentially dormant, the birth incomplete, in effect.[3]

In the ensuing days or weeks following birth, nature found ways to compensate for some of the isolated infant's needs, other ways to turn on the visual system, for instance, and make up for the failure in procedures she had designed. A compensatory operation, however, is always more indirect, slower, less effective, and far more expensive in growth energy than the innate model-response dynamic. Many of nature's associative functions are compromised when nature must compensate—and we have forced her to compensate in many developmental areas.

Statistically, infants deprived of early face stimuli and all the attendant benefits showed no signs of visual awareness or consciousness until ten to twelve weeks after birth. This contrasts sharply with the two to three minutes it takes to display these capacities when nature's model imperative is met. Masked faces, bright lights, and a drugged mother and infant simply don't provide audiovisual communication, number one in the family triad of needs. Failure of nurturing in general often follows. Compromised to varying extents when this first critical period of development is missed, a compensating vi-

---

3. The astonishing damage done by medical childbirth is probably the most exhaustively studied research item in history. In fact, the quantity and conclusions of this research rival the medical evidence against smoking. And if you think tobacco companies have been demonic in their radical disregard of human life, medical childbirth has been far more insidious and destructive. Further, because we have made a religion of medicine and sacralized our medicine men, the medical industry has far greater political and legal clout than the tobacco companies—which means we find it far easier to follow the medical lead and demonize tobacco men. Wonderful ironies.

sual and sensory system will not function to the maximum intended by nature.

I am aware that significant change began to take place in hospital birthing practice from the early 1990s. A whole group of us had been actively working for this on many levels since the early 1970s. Institutions are notoriously resistant to change, of course, and the medical world is hardly an exception. But the changes in birthing have been sporadic and limited, not widespread, as often claimed. Some hospitals have changed some procedures on behalf of some patients, generally those of higher income and education and those who demand it, but many hospitals still function on the old, antiquated standards. This is particularly the case for minority mothers and the inner-city poor, who are far more apt to be uninsured and thus of no profit to doctors or hospitals. Such women receive minimal care all too often, as a massive study by the Department of Health, Education, and Welfare showed years ago—and minority mothers have suddenly emerged as making up almost 50 percent of all birthing patients, a bit too large a proportion of the populace to ignore any longer.

When we look at the mounting crisis in the lives of young people today—the crises in family, education, social structures, deteriorating health and well-being, increasing violence in all its forms—all spilling over into the adult world in ever greater quantities, we must factor in our long century of disruptions of natural process on every level, starting with childbirth, bonding, and early nurturing. Our intellectual high brain can rationalize whole volumes of reasons and causes for our mounting disease, but our ancient brains, the foundations on which we stand, are subject to natural process unadorned and have no access to our rationalizations for breakdown as substitute for function.

On many levels contemporary life undermines the family triad of imperatives for development that Paul MacLean so clearly articulated. In so doing, the way opens ever wider for enculturation on ever more stringent levels. The results are not encouraging.

THE NEW INDIFFERENCE

▼

Back in the late 1960s, professors at the University of Tübingen, Germany, noticed a serious drop in sensory perception and general awareness in their students. (The same drop was noted in 1966 in the United States.) Students didn't appear to be as aware of information from their environment or schooling or didn't seem to register it as young people had previously. A corresponding deterioration in learning patterns was also evident. The German Psychological Association joined the university in a research project to determine if such a shift could be quantified. Tests involving some four thousand test subjects—young people in their late teens to early twenties—were carried out over a twenty-year period. The conclusions can be summarized thus: "Our sensitivity to stimuli is decreasing at a rate of about 1 percent per year. Delicate sensations are simply being filtered out of our consciousness." In order for our brains to register it, ". . . especially strong stimuli" are required. (The translation of the German reads that in order for our brains to register it, "brutal thrill" stimuli are necessary.)

Most noticeable was this elevation of what is termed the *gating level* of the ancient RAS, or reticular activating system, where sensory input from the body is collected, collated, synthesized into basic world information, and sent up to the higher brain centers for processing. The high-intensity stimuli to which these young people were subjected from birth along with the corresponding lack of appropriate nurturing and natural development resulted in a high level of stimuli that must be received in order for cognizance to form. Sensory information below a certain level of intensity or weight was not registered because it was not of sufficient strength to cross the high RAS threshold into conscious awareness and perception.

Dr. Harald Rau, of the Institute of Medical Psychology at the University of Tübingen, said,

> It is apparent that the cross-linkages [networks for sensory synthesis and associative thinking] have been reduced, and that the capacity [to screen out stimuli] has been enormously increased using direct stimulus carriers working parallel to each other. . . . Previously, an optical stimulus would be directed through various brain centers and would also activate the olfactory center, for example. Today it appears that entire brain areas are being skipped over. The optical stimulus goes directly and exclusively to the visual center . . . the stimuli are then processed faster, but the stimuli are inadequately networked [not integrated by other stimulus centers] and not enhanced with emotional input.

There is no affect—no emotional intelligence. Information is processed without evaluation, thus without reference to areas of knowledge or meaning and without emotional response. The claim of the research people is that those born before 1949 show "old-brain" reactions—that is, the norm of the time. Those born between 1949 and 1969 show modified brain action. Those born after 1969 show new-brain functioning. The new brain can tolerate extremes of dissonance or discord. In a perceptual process that would otherwise be harmonious, disruptive and inappropriate stimuli are processed without the individual noticing the discrepancies. Gert Gerken notes that new-brain people have "grown up with contradictions and they can handle them. That which used to produce a split or division of consciousness, today is the norm." Gerken refers to the "new indifference," the mental ability to unite elements that are not logically related and the failure to recognize severe logical fallacies—which results in a young person meeting everything with equal indifference. Because the brain can't bring contradictory pieces of information into any kind of relationship, it treats everything with a relative uniformity of low-grade response.

Consciousness is becoming more restricted, the research claims—the brain processes more intense levels of information and less of it reaches our consciousness. The brain has always adapted to changes in its environment by changing its own organization. "But now . . . our brain is not adapting. It is rebelling against the world and changing [the world experienced] by changing itself."

The studies show that enjoyment and aesthetic levels have dropped dramatically. Fifteen years ago people could distinguish 300,000 sounds; today many children can't go beyond 100,000 and the average is 180,000. Twenty years ago the average subject could detect 350 different shades of a particular color. Today the number is 130. The brain "loses its standards and degenerates into a kind of dialectic processing of sense impressions. . . . The brain stores opposing and contradictory information without creating a synthesis."

These young people must have a steady input of high-level stimuli or else sink into sensory isolation and anxiety. Natural settings such as parks and rural areas are avoided because they don't offer sensory input intense enough to keep awareness functioning. German psychologists have speculated that a generation with such changed brains will create an environment of such intense stimuli that a normal brain might not survive.

As a means of comparison, the total sound level of a preliterate jungle society is about that of a modern refrigerator.

WHAT "FORTUNATE" CHILDREN LACK

▼

In my book *Evolution's End,* I related Marcia Mikulak's research on sensory registration in children in the mid-1980s. Ashley Montagu and French physician Alfred Tomatis had both reported on our failure to physically nurture infants through touch, leading to increasing sensory deprivation and neural impairment. Mikulak, an independent child psychologist, employed standard Gessel tests to determine the level of a child's sensory awareness, eventually devising more extensive tests of her own. She examined young children from a wide range of cultures—from the preliterate societies of Brazil, Guatemala, and Africa, to the highly literate countries of Europe and America—and found that the children from primitive settings averaged levels of sensory sensitivity and conscious awareness of their surroundings that were 25 to 30 percent higher than those of the children of industrial-technological countries. Preliterate children were more aware of what was taking place among the people around them and what was said to them and asked of them, as well as the general sights, smells, tastes, and touches of daily life. They knew the names and characteristics of the flora and fauna in their environment, which few if any of our industrialized children or adults do. Mikulak's studies were ignored. Those of Tübingen and the German Psychological Alliance, published in 1995, have equally been ignored.

In *Evolution's End,* I also quoted from studies made in the late 1980s of the learning ability of children in so-called primitive groups such as those in Guatemala and similar countries that have severely low standards of living. When these "deprived" children were put into a learning situation equal to those provided for our well-cared-for children, the deprived children showed a three to four times higher learning capacity, rate of attention, and comprehension and retention than our "fortunate" children. Deprived of advanced electronics, these primitive children were given the most necessary things—love and nurturing—and they played continually and developed to the maximum their society afforded.

WORD DETERIORATION

▼

American high school students of 1950 had a working vocabulary averaging 25,000 words. Today that level is 10,000. As of 1998 some 85 percent of all academic honors in the United States were taken by foreign-born students. Offspring of these students may, in turn, keep our standards from disappearing for perhaps one or two generations more, but that will be all. Sooner or later they will become we, and who will be left to comprehend that intelligence itself has deteriorated?

All we have done in response to this astonishing cascade of breakdown, besides building more prisons, is increase to well over a million the number of daily doses of Ritalin and a whole family of pharmaceuticals designed to alter the behavior of children. Meanwhile, we legislate for stricter learning standards, "getting tough with kids," increasing homework and testing. It's interesting to note that testing is interpreted by all of us as a judgmental threat and shifts our energy and attention from the emotional-cognitive brain and prefrontals to the R-system, which compromises whatever higher intellect we may have. In fact, being back in school to take a test is a common nightmare of adults.

For thirty years I have made the unpopular proposal that our treatment of our children has made them increasingly uneducable by the time they reach school age. Mark, then, a further prophecy, made by a score of better heads than mine, that computerizing schools will bring this whole mounting chaos to its terrible, irreversible conclusion. Age-inappropriate use of electronic devices undermines the very value of those devices.

In the remainder of part 2 we will explore a few more cultural issues to fill the objective of this book—to examine the twin phenomena of violence and transcendence. Perhaps through this we may avoid the depths of the former and reach the heights of the latter. Unearthing the roots of violence is not a comfortable process—but they must be revealed if we are to wake up, rise, and reach beyond them.

SIX

▼

BIOCULTURE AND THE MODEL IMPERATIVE

Culture is a body of knowledge concerning survival in a hostile world, inherited and passed on from generation to generation.

—GRETCHEN VOGEL*

Research published in 1998 provides a clue to our evolution and development, and perhaps to the slowly swinging cycles of civilization. This research concerns brain growth during gestation and, in addressing its subject, manages to cast a light that illuminates our current personal and social dilemmas. Before looking more closely at this study, however, it would be helpful to review briefly some facts.

Instinctual patterns for reproduction, birth, and infant nurturing are inherited from our ancient mammalian ancestry. Nature passed on to us not just the rudiments of the larger and more powerful neocortex, but also the rudiments for those far larger and lovelier breasts of our mothers as well as a full program for employing those marvelous mammaries to best advantage: nurturing offspring.

Mother and infant are designed to be a dynamic that activates in each nature's agenda for nurturing and well-being. The infant unlocks in the mother a wisdom and knowledge gained over eons as the mother unlocks in the infant the intelligence to be fully human and, eventually, to nurture his or her own offspring in turn. The mother's influence is far more pervasive than she may be aware of. The research report mentioned above can be

---

*Paraphrase from Gretchen Vogel, "Chimps in the Wild Show Stirrings of Culture," *Science* 284, June 25, 1999, 2070–2073.

summarized thus: If a pregnant animal is subjected to a hostile, competitive, anxiety-producing environment, she will give birth to an infant with an enlarged hindbrain, an enlarged body and musculature, and a reduced forebrain. The opposite is equally true: If the mother is in a secure, harmonious, stress-free, nurturing environment during gestation, she will produce an infant with an enlarged forebrain, reduced hindbrain, and a smaller body.

The oldest evolutionary brain in our head (and body), you recall—the reptilian or hindbrain—provides for fast physical reflexes; is geared to brute strength driven by primary survival instincts hardwired for defense; and is reflexive, not reflective and not very negotiable. The forebrain, on the other hand, gives rise to our intellectual, verbal, and creative mind, functions more slowly, is reflective, and is far more intelligent and negotiable than the defensive, hair-triggered, and reflexive hindbrain.

In her evolution, nature didn't add a forebrain with its reflective, creative intelligence until she had worked out the logistics of a protective, survival-oriented brain upon which she could build her new one. So nature's shift in uterine brain growth toward the kind of environment that a new life must deal with follows an established, adaptive common sense that would please the most ardent Darwinian. Note, however, that nature shifts from an emphasis on physical survival to an emphasis on intellectual enhancement whenever she gets the chance. That is, she moves for a bigger forebrain at each opportunity, asking in effect, at each conception, can we move for greater intelligence this time, or must we protect ourselves again? This is, after all, an organic and most intelligent life process, not a rote chemical mechanism. Perhaps at times of catastrophe our general brain structure suffers a setback, but because evolution obviously moves toward higher forms of intelligence, nature can recoup quickly whenever the environment is favorable, responding even to individual cases and the internal environment of just one mother.

THE BIOCULTURAL DYNAMIC

▼

For years Bruce Lipton and other enlightened biologists have observed that environment influences genetic coding every bit as much as conventionally recognized hereditary factors. Lipton found that from the simplest cell on up, a new life unfolds in one of two ways: It can either defend itself against a hostile environment or open, expand, and embrace its world. It can't do both at the same time, however, and environment is the final determinant in the decision.

That neural growth will shift from a defensive, combative stance to one that is reflective and intellectual—or vice versa, according to the mother's emotional state—offers us the chance to make a profound shift in our history and to take our evolution in hand. Even in the middle of pregnancy, if there is a change from negative to positive in the mother's emotional life, the direction in fetal brain growth changes accordingly.

That a mother in a safe space produces a strikingly different brain and child physiology than one who is anxious clearly illustrates nature's model imperative. The mother is the model of the eventual child on every level and a new life must shape according to the general models life itself affords. For, as is true in all cases of nature's model imperative, the character, nature, and quality of the model determine to an indeterminable extent the character, nature, and quality of the new intelligence that manifests.[1]

This all indicates a biocultural dynamic—our biology influences our culture and our culture influences our biology. A sufficient number of children born predisposed toward defensiveness and quick reflexive survival reactions will tend to change the nature of the society in which they grow up. Recall the German studies on indifference in young people that were cited in the last chapter. To protect itself from such reactive people, the society will become more defensive and wary itself, creating the very conditions that bring about more people with larger hindbrains and smaller intellectual-creative forebrains. Thus angry, defensive people tend to reinforce their condition in the next generation, civilized society disappears, and a culture is born that grows more and more explosive and dangerous with each generation.

Culture has been our principal environment of mind for many millen-

---

1. The neural pruning right before birth can now be seen in its true light. Nature provides her usual overproduction of neurons in utero to cover either a greater hindbrain or a greater forebrain, and so concludes gestation with some 30 percent more neural material than is needed to meet the upcoming environment. The leftovers will be from either system according to this practical selection, and must be pruned right before birth to make way for the growth spurt to come at birth itself. This growth spurt at birth is preparation for the infant's adaptation to his new environment.

nia, and through the dynamic of culture and biology, humanity fell into a vicious cycle long ago, a trap from which only the prefrontal-heart dynamic can deliver us. Nature has continually offered us this escape, but, time and again, circumstances breeding fear in us have turned her down.

THE EFFECTS OF THERA AND VENUS

▼

Maria Colavito, in her biocultural studies, described a hypothetical shock that the reported blowup of the hypothetical Mediterranean island Thera brought to the world some time before the first millennium B.C. She proposed that this terrorized the survivors and disrupted the earth-trusting mind-set of our species, leading to the Phoenician alphabet and a long parade of harsh patriarchal cultures and religions. One is hard put to account for peoples of the whole globe being affected by this rather enclosed, comparatively localized affair until we connect this possibility with the theory presented by the scientist Velikovsky back in the 1950s.

He proposed that around 1500 B.C. the planet Venus was caught in the sun's gravity field and brought into our solar system between Mars and Earth. This fiery interloper temporarily disrupted the orbit of Mars and the spin of all nearby planets as it found its own orbit. On the day of Venus's arrival, total havoc would surely have reigned over the whole of Earth. Perhaps, then, the arrival of Venus and the destruction of Thera may have coincided. Venus's earthshaking intrusion would have brought Thera (and no doubt many other volcanoes) to thunderous eruption. We have only to consider the serious atmospheric disturbances following the 1888 eruption and subsequent destruction of the island of Krakatoa in the Pacific.

Velikovsky was excommunicated by the scientific community of his time for publishing his theory as a popular book (and making money on it) rather than submitting it to the conventional peer reviews of scientific journals (where it could have died quietly of neglect). Further cause for scientific dismay was that he had concocted the notion from passages in the Old Testament and other ancient texts and proposed that the Old Testament had historical value.

Velikovsky stated that Venus had a retrograde spin (opposite that of all the other planets) and that its spin would be slowing down and eventually would of necessity reverse and spin in sync with all the other planets. (That they have the same orbital direction determines that the planets of our solar system have the same clockwise spin.) He also predicted, among a number of things, that the atmosphere of Venus would be intensely hot. Virtually all of his predictions of what we would discover about Venus have been borne out in subsequent space probes and studies, and the Academy of Sciences was forced to acknowledge posthumously the accuracy of Velikovsky's predictions and grudgingly apologize for the rude behavior of its members. Of course, it promptly forgot the incident and Velikovsky's observations.

I surely do not base my thesis here on the risky premises of Velikovsky or Colavito, but some such catastrophe may well have taken place that left survivors with good reason never again to trust the Great Mother as in the past. Perhaps through trauma our species suffered a setback, breeding an intellect based on distrust, fear, and the attempt to predict and control a hostile natural world in the interest of protecting against it, which ultimately produced culture as we know it today.

CULTURE AS A FIELD EFFECT

▼

In chapters 3 and 4 I defined and defended the hypothetical concept of fields of intelligence, causal forces, or influences functioning in dynamic with the neural fields of our brains and used examples from the idiot-savant phenomenon to provide evidence that field formations are brought about and sustained through human experience itself. By the very nature of the human brain we create field effects and are affected by them. In this way, fields become culturally shared and move history accordingly. One of the largest factors in our history, perhaps making that history what it has been, is that culture is itself a field, independent of any of its expressions.

According to recent anthropological research, culture is a body of knowledge concerning learned survival strategies that are passed on to our young through teaching and modeling. (Anthropologists claim to find traces of culture even in the higher apes.) It becomes the living repository of our species' survival ideation and is at the root of every issue of survival.

Although at some very early point in prehistory the focus of our survival may have centered on the saber-toothed tiger or his equivalent, for many millennia now the focus of our survival has of necessity centered on culture itself, a fact that is difficult to grasp but one I will try to clarify, for surviving our own violence is a growing issue and culture and violence are intertwined.

Accept for sake of discussion this definition of culture as an aggregate of ideas about survival, a taxonomy that lifts disparate notions into a coherent and powerful whole. Culture as a field effect is thus inviolable, its contents or expressions interchangeable and even incidental because culture absorbs and transforms any content into its own formative structure. Similarly, anxiety is a state of chronic, free-floating fear—fear without an object. Such a state acts as a catalyst, changing every object, every event into its target, making an event fearful whether or not it deserves to be considered so. Anxiety can become the lens through which we interpret our ongoing experience.

Culture, then, is a mutually shared anxiety state, a powerful catalyst of thought that converts all events to its own nature. Mathematics as a field contains mathematical content, music as a field contains its possible sounds or styles. Culture has no analogous content but is rather a pattern of survival behaviors that, once learned, expresses in any content or context, adapts every event to its fearful nature, and colors every aspect of our life.

Arising as a set of beliefs and practices centering on physical survival, culture breeds a mind-set centered in our ancient hindbrain, nature's survival system and means for defense. As pointed out earlier, once conditioned, our ancient R-system's patterns are as nonnegotiable a reflex as jerking our hand from a hot object or closing our eyes if it seems an object will hit them. In just this way any sensory report resonant with this ancient encoding activates the whole neural pattern of defense. Like anxiety, culture embraces every negative idea or possibility as its own until all that's embraced in turn embraces.

Long ago, wisdom dictated that we hand down our fears of saber tooth along with the defensive procedures protecting us from him. Eventually, however, with the tiger long gone, we were left with only that defensive inheritance as an empty slot or category, a mind-set that colors all reality, a movable feast of anxiety. Although our children are able to absorb this attitude from the very air they breathe, we have, since earliest times, made sure they emulate such learning. Over time, formed as the means to defend against predators that were no longer a threat, culture created the very conditions for which it was designed. Its defensive procedures became necessary for protection against human predators shaped according to those very procedures, hybrid humans operating from their defense system and using their neocortex to enhance those defensive procedures. Once set in motion and locked into our ancient reptilian brain and its hardwired survival memories, this cultural effect reproduces itself automatically and is thus passed on.

Our greatest fear, the late philosopher Suzanne Langer said, "is of a collapse into chaos should our ideation fail us." Culture is that ideation, or set of ideas. The foundation and framework of our worldview, self-image, mind-set, faith, and belief are culturally determined. Our grounding in culture and culture's grounding in survival are so intricately a part of our mental fabric that such roots are seldom if ever exposed, and even then can hardly be recognized for what they are. Culture is the mental environment to which we must adapt if we are to survive, and in our adaptation and survival we automatically sustain culture.

Enculturation, culture's imprint on us from the time of conception, makes Langer's collapse of culture's ideation a virtual impossibility. Our survival ideation determines the very shape of our brains and the neural fields within them, as this chapter's opening research indicated. All internal and external facets of our life reflect culture, an emotional environment that determines the content by which our genetic blueprints are filled from the beginning. Threaten our current cultural body of knowledge and you threaten our personal identities, our core being. Such a threat can lead us to behaviors that go against survival—at least for the victims of our reaction.

Gil Bailie and the French philosopher René Girard discuss a direct relationship between culture as a function and violence as its means for sustenance. Thus Thomas Jefferson's famous statement that the tree of liberty must be periodically watered with the blood of tyrants and patriots embodies and ennobles the cultural effect. Likewise, A. E. Houseman's dying young soldier in *A Shropshire Lad* says, when he hears the living men shout, "God save the Queen": "O God will save her, fear you not, / be you the men you've been, / get you the sons your fathers got, / and God will save the Queen."

A culture's violence directed against neighboring cultures acts as a cultural emetic, purgative, and restorative. Nothing pulls together a disparate and dangerously enculturated group into a cohesive unit so quickly as a good war. Bailie and Girard analyze and describe the cycles of periodic warfare and murder through which culture has sustained itself for untold millennia. Having a clear enemy to demonize provides a clarified target for the free-floating anxiety and its accompanying fear and rage that enculturation brings. But when that restorative solution of war begins to break down, no longer delivering the necessary fix, as it has in our time, the violence bred by culture begins to turn inward, leading to self-destruction.

Our need for scapegoats is palpable but produces less and less effect with each passing year. Cultural anthropologist Leslie White proposed that cultures are born into history to run through their cycle and die. Further, he pointed out that they all die by their own hand. Through demonizing our enemies, then, we stave off cultural suicide.

CULTURE AS ARCHETYPE

▼

Carl Jung referred to a demonic archetype that had long percolated beneath the surface of Germanic European thought. This archetype, galvanized and brought to action through the Nazis, was used to mold a demoralized people into a cohesive unit. In another context Jung referred to the danger of the ego being "inflated by an archetype." Culture can become a kind of psychic entity that can possess and/or inflate a person or even an entire country and achieve its violent ends through such possession and inflation.

Like the mathematical field that can solve vastly complex arithmetical problems and present them through the resonant neural field of a savant, culture is a field effect built up to tremendous dimensions over the millennia, a form of semi-sentient intelligence functioning as a pseudo-universal force. It expresses in infinitely various ways through myriad brain structures that are born into it and are shaped in conformity with it. In this way the various types of culture never change the cultural effect. African-American culture, Anglo-American culture, Latin American culture, Native American culture—all of these are simply culture wearing any face available, and there is no end to the faces culture can make.

ENCULTURATION AND SOCIALIZATION

▼

The terms *culture* and *society* present semantic problems. By implication culture includes the highest achievements of humankind—art, music, philosophy, science, astronomy. To some people society implies a small, exclusive class of people feeding off the social body and "living high" while, perhaps, patronizing the arts. Blake, as a poet and artist, scorned society and praised the cultural life that gave us poetry and art. Elkhonon Goldberg's term *civilized mind* offers a possible alternative to both *culture* and *society*, but even so, the terms *civilized* and *uncivilized* veil the fact that enlightened societies breed atom bombs and holocausts, while jungle societies may live in relative peace and harmony. Blake once commented that humans weren't made for the industrialized city nor the jungle, but the garden, nature transformed by man. Perhaps the word *civility* would serve, but the anthropologists' use of culture as a survival orientation tipped the scales in my decision. In the following text I use *socialization* as a definition of *civility*, that which brings us together in cooperative benevolence and nurturing, and use *culture* as a shared conglomerate of survival strategies that breed group violence and despair.

Socialization in this sense is instinctual, while culture is not. Our social impulse arises from the so-called herd instinct inherited from our mammalian ancestors. The pleasure in gathering together with our own kind, found in most mammalian and avian life, is the source of community and fosters the model imperative; extended nurturing and care; mutual sharing of aesthetics, events, dreams, hopes, ideas, and ideals; mutual appreciation of works, skills, creativity, cooperative ventures; and the sharing of the higher, broader expanses of love—love of neighbor, self, and God. The possibility of being the last person on earth is a science-fiction nightmare that preys on our instinctual drive to socialize. Even though such a scenario would mean that we could own the whole world, it would be an empty ownership. Socialization amounts to relationship and sharing, the very juice of life. Northrop Frye suggests that the phrase "the kingdom of heaven has come among you" actually means it comes about through our relationships with each other.

The needs of our species' gene pool are met through this herding instinct, but the gene pool is not the primary impetus for socialization, as our enculturation would have us believe. Consider how in nearly every species males tend to gather with males, females with females (as at a cocktail party), except at specific mating periods (such as after the party). This birds-of-a-feather tendency springs from the simple pleasure we find in the company of our kind.

Enculturation, on the other hand, is not instinctual but instead the result of conditioning, our enforced learning and adoption of ideas about survival, including techniques believed necessary in our particular cultural environment in order to survive. Our imitative monkey-see, monkey-do compulsions actually arise from our oldest reptilian brain system, which is linked to survival and fight-or-flight injunctions of the old mammalian brain. Ironically, this combination provides the principal tools employed in enculturating our children. Enculturation is not instinctual; we must capitalize on and use our survival instinct to bring it about. With regard to enculturating our children, lacking all conviction otherwise, we move with total, passionate intensity. Convinced we must pass on this survival knowledge, we pound it into our offspring "for their own good" as it was pounded into us for our own good. Schooling is treated in a similar fashion—no matter how much pain schooling may have caused us, to save our sanity over having lost the richest, loveliest years of our life to the process, we rationalize that it must have been good for us! And we then subject our children to it in turn; they prove our point by becoming like us, confirming our worldview, joining our mass anxiety, and verifying it by coming on board. We have very little choice in the matter, but hope springs eternal that this time we will make schooling work.

It never has.

OUR CHILDREN'S GROWTH: JOYFUL LEARNING OR CULTURAL CONDITIONING?

▼

A child's socialization, which can be characterized as learning in its most complete form, encouraging reflective thought, is instinctual and arises spontaneously on its own. Culture is something quite opposite: an intellectual, arbitrary conditioning and enhancement of automatic reflexes that must be both induced and enforced. A society—the product of socialization—is made of spontaneous nurturing and love, while culture can bring quiet hate, which can lead, sooner or later, to a child's subtle or flagrant rebellion. Such rebellions are forcibly put down through the infliction of pain, fear, guilt, and shame, or, if none of these works, then through isolation, exclusion from the group, or the labeling of the rebellious child as dysfunctional or unfit.

These two extremes as they relate to our children's growth seem clearly represented in two phrases I have read: *Living Joyfully with Children* is the title of a splendid little book written by Bill and Win Sweet, which contrasts with the words I once saw on a huge highway billboard stating that raising a child is the toughest, hardest task we ever undertake. Now that's good, practical, tough-minded advice! Forget that joyful nonsense, lest culture disappear.

Infants instinctively resist enculturation because they intuitively sense in it a denial of life that robs us of our spirit and our loving, willing, thinking being, as Ludwig Feuerbach expressed it. This is a subject explored at length in chapter 7. Resistance is futile, however, for it ultimately brings about the use of intellectually derived techniques to overcome resistance, as implied by the billboard message I saw on the road. Many parenting books focus on how best to enculturate your child, carefully cloaking advice with the current politically correct phrasing and playing on parents' concerns over the child's education, place in society, career, fame, and fortune, and constant threat of failure to achieve these.

Without exception, these cultural techniques involve carefully masked threats that prey upon the child's rapidly learned fear of pain, harm, or deprivation, and more primal anxiety over separation or alienation from parent, caregiver, and society. No matter how we camouflage our intent both to ourselves and to our child, most parenting and education (except, perhaps, Waldorf and the best of the Montessoris) are based on "Do this or you will suffer the consequences." This threat, in fact, underlies every facet of our life, from our first potty training through university exams, doctoral candidate orals, employment papers, income tax, on and on ad infinitum down to official death certificates and burial permissions, no matter how high on the cultural totem we climb. Culture is a massive exercise in restraint, inhibition, and curtailment of joy on behalf of pseudo-safety and grim necessities. We live out our lives in the long shadows it casts.

Because enculturation is both induced and sustained by threatening us with possible harm, deprivation, or even death, as from accident or illness, from the beginning of life fear becomes the foundation of our mind-set, leading us unconsciously to screen every event for its potential threat and interpret its nature accordingly. Such cautious directives continually activate our instincts of defense, which enculturation plays upon so well.

Ironically, but more seriously, enculturation is enforced by threatening us with a loss of our true and naturally spontaneous sociability and desire for relationship. Loss of sociability translates as loss of even the chance to love and be loved, which amounts to a living death. Because culture must guarantee the pursuit of happiness—happiness is our birthright to love and be loved—it offers counterfeits such as illusions of possible fame, fortune, and safety as our birthrights in order to keep us on its treadmill. Culture never allows happiness to be achieved because in doing so it would guarantee its own disappearance.

Paradoxically again, our cultural ideation, once centered on fears of saber tooth, is now focused not on survival in the natural world but on survival in culture itself. We deliberately enculturate our children to protect them from culture, though this is never spelled out, seldom recognized, and a serious offense to our sensibilities should it be pointed out. How many times do we hear parents, reflecting on their child's future, ruefully point out "Man! It's a jungle out there!" The jungle is culture, the predators enculturated humans. Oddly enough, culturally engendered prescriptions for child rearing create a new generation of people chained by culture and compelled to spend a lifetime correcting its failings and mediating the pain it deals through restriction, never noting that with each correction and mediation, culture is strengthened. "Never before has a generation faced the challenges you young people face," the commencement speaker intones, generation after generation. A lifetime spent taking up this challenge wins culture's applause and Nobels—and helps perpetuate it.

The summary statement of enculturation—and the clarification of its deadly opposition to the gospel—is, "Ask not what your country can do for you; ask what you can do for your country." That this frequently quoted statement was made by a Christian president in a Christian country points up the fundamental antagonism between the gospel and state-religion supposedly based on it. In the gospel we are told that, "The Sabbath was made for man, and not man for the Sabbath," which is the fundamental cultural, religious, and legal heresy of the ages by default. Were the gospel heard or even this one quote from it comprehended, culture and its states, religions, and law would disappear and society could emerge as our natural state.

Culture is the fundamental deviancy of intellect from intelligence, and because of its massively unnatural, arbitrary, and illogical nature, it requires an equally massive energy to sustain it. Without periodically imploding, a culture's energy needs would consume the whole world. In fact, while concupiscence, an uncontrolled appetite for sensual consumption beyond all restraint, is the classical Christian sin, its contemporary implications go far beyond sexuality. Consumerism is the modern concupiscence, and a bottomless pit. Yet our great model spoke of "life more abundant." Today we interpret that as more goods on our shelves, more food on the table, while he was speaking of the lifting of restraints on our spirit—the restraints of states and religions, for example, which create laws and, as handmaidens of culture, make war. Law and war, religion and science: These are the ultimate expressions of our restraint of one another and our spirit.

The nature or character of a myth or religion is incidental to the force of the culture, which both embodies and gives rise to myths and religions. And abandoning one myth or religion to embrace another has no effect on culture, which produces myth and religion automatically. Science supposedly supplanted religion, but simply became a new religious form, an even more powerful cultural support, and an equal source of restraint on our spirit.

COUNTERFEITS OF TRANSCENDENCE

▼

A new and all-pervasive negative field has been growing among people worldwide, an angst or fear without an object and tinged with rage. The angst is fed or fueled by mass media. Saturating all societies our mass media feed into and feed on, this global angst is a typical biocultural process. No one knows where it might lead. Already it is a kind of demonic spirit that blows where it will.

This angst-ridden energy is nothing less than our longing for transcendence, which, in light of its enormous evolutionary power, must be derailed or subverted by culture if culture is to survive.

But is culture real? Or is it, like a Tibetan *tulpa*, a phantom of the human intellect? Once isolated from the intelligence of the heart, once entrained with and by culture, we interpret cultural survival as our own personal survival and respond as . . . culture. We, then, are culture, just as we are nature and evolution.

Delivery from this massive and ancient error of mind has been the intent of every great being of history, and was surely the intent of Jesus. Tackling culture was the thrust behind the cross. Jesus demonstrated that our true nature is transcendence itself, and his attempt to awaken us to enculturation and its power strikes me as the most outlandish tilting at windmills in history.

Confronting culture makes the mythical struggle of Prometheus seem like child's play. Small wonder we mythologized Jesus in the manner of Prometheus, for this is our culture's common ploy for neutering any viable threat. And what is the great threat of beings like Jesus? Pointing out the illusion of culture and the reality of our transcendent nature.

In the myth of Prometheus, fire, a secret of the gods, is stolen by a godlike man. The human wasn't smart enough to make fire on his own without the gods' help, so it became necessary to steal from "out there," from cloud nine, what was actually humankind's birthright. Likewise, the myth attached to the cross made transcendence a secret of the gods that was also stolen for us by a mythical god-man—one owned by a church and religion based more on the myth than on the actions at the core of the myth. Thus the gospel has become a commodity that any smart cultural culprit can profit from.

Yet in the dark hours of night, this issue of the cross and culture awakens me with the ponderous enormity of that all-too-real undertaking, still unfolding and unresolved today.

SEVEN

THE
ENCULTURATED
SELF

All evil consists of self restraint or restraint of others.
All evil acts are murderous.

—WILLIAM BLAKE

William Blake said the only sin was the accusation of sin. Accusation, in any of its forms, is a negative judgment, and a negative judgment in any form ruptures relationship—the classical definition of sin. Christianity narrowed this to the relationship between human and God, while the gospel shows that the target of a negative judgment is incidental; who is doing what to whom to rupture which is mere intellectual froth. The heart of the issue is negative judgment—which I shall refer to simply as judgment—as an act of mind.

Being judged by someone offends us if the judgment is true and more so if it is false. When we accuse or judge another, it has the same effect on us as being judged ourselves. Any judgment we make, no matter of whom, registers in the heart as a disruption of relationship, and the heart dutifully responds on behalf of our defense, shifting neural, hormonal, and electromagnetic systems from relational to defensive. Our sensory system reflects those shifts in its source material and the environment we experience changes accordingly, although perceived as the usual natural phenomena of our world to which we respond as usual. Creator and created are giving rise to each other, we are judged as we have judged. Sowing with arrogance, reaping with tears.

If we examine our roof-brain chatter or stream of consciousness, that nonstop flow of thoughts in our head, we will find that it arises as naturally as breathing and centers almost exclusively on judgmental, accusatory fault finding: Someone or some event has offended us, threatened us, failed to meet our lofty standards or probably will in the future. Because this train of thought seems almost cellular in origin, beneath our volition, why is its content so often negative?

THE POWER OF THE NEGATIVE

▼

Blurton Jones, one of Nobel winner Nikos Tinbergen's group of ethologists in England (ethologists study animal behavior), gave a clue to the enigma of negative streams of consciousness years ago in his research on the pointing syndrome. All mammalian young are genetically driven to interact with the objects and events of their environment, upon which they build their neural imprints. Any new, unfamiliar object or event powerfully signals our young to interact with it to build such a structure of knowledge. As a rule, in their initial encounters with their environment, infant animals check for their mother's okay, which she gives through a variety of subtle sensory cueing, before they interact with a new phenomenon.

In the nest or home all objects and events are safe for interaction, but in the great outdoors, caution is the rule. Our toddler points to something unknown and checks his caregiver's response. If positive, the toddler follows through with a complete sensory inventory of that phenomenon, tasting, touching, smelling, listening, and talking to it, in order to build from it a structure of knowledge. Such imprints include the name, if given, and the emotional state experienced during the exploration. Thus, the world the child constructs will be one shared with the mother.

Seldom will a young wild creature disregard a mother's cues that an object or event might be dangerous. Such warning is the primary signal on which mammalian life has depended throughout history. In our evolutionary past a child disregarding the danger signs of a caregiver was tiger's lunch and left no progeny. To us, the progeny of those who heeded warnings back then, obeying a warning is still one of our strongest instincts, encoded by our ancient sensory-motor and emotional-cognitive systems. Should a parent's directives be ignored, that parent's own encoded survival signal fires and can cause upset or anger on the part of the parent. Survival is no joke.

In addition to communicating silently with body signals, most adult animals have a repertoire of warning calls that can also alert the young that a saber-toothed tiger might be around. In our case, language has largely supplanted warning calls. NO! can replace the whole complex of signaling, and, if ignored, is generally followed by reprisal to teach the child that "we mean it." NO! therefore not only triggers the same ancient reaction our ancestors had to saber tooth, but also may indicate mother's wrath, which is an even greater threat. Abandonment by the mother is the greatest of all a child's fears, tantamount, even, to death.

NO! and any subsequent negative event that results from ignoring it trigger the fight-or-flight response from our saber-toothed-tiger days as translated through our ancient amygdala.[1] If we give an emphatic, harsh NO! our poor dog hangs its head and curls its tail between its legs, seeking our forgiveness with a look of crushed sadness and shame on its face. Our animal brains are shared systems and never forget.

Now, all objects and events are fair game for a toddler's exploration in the haven of home because home is an extension of mother, and mother is the safe space itself. In infant-mother dynamics, a mother's actions are automatically models for an infant's action, whether in the safe space of the nest or outside it. In a supermarket, for instance, mother handles plenty of those thousands of objects on the shelves surrounding her, so such objects are obviously fair game for interaction and the child quite naturally follows suit.

Problems arise, however, when the child follows his genetic encoding and explores an unknown in the safe space of home but meets with an emphatic NO! or DON'T! from the caregiver. What was automatically safe to do seems suddenly and arbitrarily not safe—a conflict of signals. Likewise, in the supermarket, when mother picks up items and puts them in the basket and toddler follows suit, an emphatic NO! DON'T! declares as dangerous what ancient encoding had just declared safe.

Monkey sees, monkey does, and monkey gets clobbered—by his very model and safe space! Thus a double directive is delivered—checkmate—

---

1. The amygdala, recall, is a critical neural module involved in memory, particularly fight-or-flight decisions, and is a kind of halfway house communicating between the reptilian brain and the mammalian brain. Virtually all memories of the first three years involve the amygdala.

and when the caregiver reinforces NO! with physical restraint or punishment, or indicates by a fierce, angry face that the relationship might end, the child understands only "abandonment." In a moment the source of all good things turns into the source of the ultimate, primal threat. That dreadful NO!, indicating saber tooth nearby, transforms the caregiver into a giver of pain and even threatens to break the bond on which life depends—and for no reason discernible to the child.

Thus NO! becomes a powerful and terrible word to the child and is generally one of the first words he speaks as he tries to get a handle on that malevolent negative force. Countering negative with negative, like fighting fire with fire, may be our first learned survival strategy. Sooner or later survival overwhelms the most rebellious will; the toddler conforms to NO!, ceases his exploration in proper fashion, and becomes one of us.

This explanation might strike you as overstated, but the youngster, caught in a serious contradiction of terms, experiences ongoing confusion, ambiguity, and uncertainty. If the safe space is no longer safe, where do we turn? Using negatives to correct behavior is at the very heart of enculturation, however, and the logic never improves. "Thou Shalt Not" is a wellspring of law and religion, the cement holding culture together, the source of all legal systems, prisons, war, and our downfall.

To the person so enculturated, however (and who isn't?), any negative meted out demands immediate attention, which is always given because such negatives trigger our ancient survival instincts. Anything resonant with survival and the amgydala alerts powerful systems of the brain and body of possible danger. Once alerted and brought into play, they will not let our fickle attention or intellect wander, but will hold us with single-minded focus so that, just as nature designed, we might do what seems necessary to survive. We as adults might laugh at the triviality of a negative alert that catches our attention, but only after making certain that it is unfounded.

As a result of such enculturation by the negative, all the news that's fit to print is generally negative news—without a negative as its foundation, or tucked in as a tidbit to induce our persistent focus, we won't pay attention to cultural process, whether it be in the form of news, television, politics, economics, ecology, health, education, religion, or any of the endless parade of follies in these worlds. The attention demanded by culture is arbitrary and contrived and is intuitively recognized by us as counter to our true well-being. But a negative enlists our attention, with or without our assent, because it registers in our primary systems as threat, and once that occurs, we're hooked.

The logic of this, or rather the lack of logic, never registers. We are not conscious of the reaction of the amygdala and its accompanying survival signals; we merely reflect the results and react accordingly. It is a knee-jerk response, like shooing a fly or scratching an itch. No media project succeeds based on "good news only" because good news doesn't trigger our alert system. Anything good indicates the safe space, the quiet background against which events can play out. The enculturated mind is cued to respond to the negative as a point of focus, which largely screens out or ignores a quiet stable base, and, because it sharpens and maintains our alert awareness, we actually begin to look for the negative.

TODDLER AT THE CROSSROAD

▼

In human development the early toddler stage is the fountainhead of cultural renewal. At stake is the activation and development of the child's sensory system and knowledge of the world, and the equally important building of his emotional-cognitive system's knowledge of what relationships with that world are like. By about the eighteenth month after birth, the child's emotional-cognitive system has formed patterns of response that will determine the nature of his relationship for life, the neural foundation of all learning. Maria Montessori claimed that "a humankind abandoned at this earliest formative period becomes the worst threat to its own survival."

Allen Schore's research shows that we all experience abandonment of a kind, which perpetuates our culture and seriously impairs our emotional-relational system itself. Recall how the emotional state of the mother determines the actual character, nature, and shape of the infant's brain in utero. Allan Schore shows how this relationship exists through the first two years after birth as well, further determining the growth, shape, and nature of the child's developing brain. One of the major growth spurts of the brain takes place after birth, and the fate of the new neural material introduced at this time is subject to the same model imperative as that introduced before birth. The way the brain is used, based on its model, is the way it forms and grows.

Schore's study concerns affect regulation, or our ability to modify or modulate initial impulses from our sensory or emotional system, and the role this plays in the organization of our self system, that unique sense we have of being an individual distinct from the world out there. Growth and development of the connections between the prefrontal lobes and the emotional-cognitive brain, with its direct connections to the heart, are what is at stake here.

TAMING THE BEAST WITHIN

▼

Schore's extensive research (which includes 2,300 citations) reflects, by default, our fundamental cultural attitudes and thinking concerning children. The first major assumption Schore reflects is that children must be socialized, a key factor in his study. By the term *socialized* psychologists mean civilized or humanized, which in turn means that a child's supposed inherent, natural animal behavior is modified by culture's arbitrarily imposed restraints. This suggests that unless his behavior is modified, the child will grow up to be, in effect, animal-like, a beast unfit for society. This myth serves to ensure our enculturation and the preservation of culture itself.

Schore proposes that the procedures we use to socialize a child play a critical role in the forming of a self system, that sense we have of being an individual distinct from our world. His notion is right, but for the wrong reasons, and has profound implications. The critical question is this: What kind of self results from such a fabric of negative assumptions?

Within the boundaries of the definition of the word on which this chapter is based, culture is our body of knowledge concerning survival—which means that the actual parent-child interaction Schore details is enculturation, not socialization. The commonly shared belief that a child must be socialized is at the heart of our own personal enculturation and resulting worldview. We have no choice but to reflect this notion, once implanted, because it registers in our ancient survival system as part of our learned survival strategies—a key factor in ensuring that we will enculturate our children as we ourselves were, and that we will probably use the same techniques that were used for us. Thus we keep culture intact.[2]

All of us know intuitively that we are not by nature savage beasts. Fewer, however, are aware that we are driven to some fairly beastly behaviors by enculturation, despite the fact that the process itself is supposed to prevent this. This irony brings us to the fundamental struggle between society and culture, which is also the struggle between intelligence and intellect, evolution and devolution, spirit and religion, gospel and myth, heart and brain, love and law, creator with created.

A CAREGIVER'S PROHIBITIONS

▼

Although the sizes of the hindbrain and forebrain are determined by the mother's emotional state while a child is in utero, the growth of the prefrontals is determined by mother-infant interactions in the first eighteen or so months after birth, and, you recall, the prefrontals are critical to all higher intelligence and to transcendence itself.

Allan Schore points out that growth and development of the prefrontals is experience-dependent, which means that the actual cellular growth and functioning of the prefrontals is dependent on appropriate stimuli from the environment. For a child in the first year and a half after birth, that environment is the mother: "Interactions with the mother directly influence the growth and assembly of the brain's structural systems that perform self-regulatory functions in the child . . . and mediate the individual's interpersonal and intra-personal processes for life."

Not only does the extent of cellular growth depend on environmental stimuli, but the character or nature of what does grow and develop is determined by the same model imperative. "The physical and social context of the developing [child] is . . . an essential substratum of the assembling

2. Research has puzzled over the lack of recall most people have for their first three years of life and the fact that functional memory with some recall doesn't begin until age three to four. This is because the amygdala, which is the module primarily involved in memory of these first three years, is fully engaged in registering our survival strategies. The hippocampus, involved in long-term memory subject to later recall, undergoes its major growth after the first three years. Thus very few of us can actually recall our survival training; we simply act it out, particularly when dealing with our own offspring.

[brain] system. . . . The tenth to eighteenth months mark the final maturation of the system in the prefrontals essential to regulation of affect [emotion or relationship] for the rest of that person's life."[3]

(This observation must be qualified based on evidence that the prefrontals undergo a major growth spurt at adolescence, a discovery not commonly known when Schore was developing his theory.)

So, with the mother present to fulfill the model imperative, the toddler learns to walk, plunging with spontaneous excitement and abandon into his exploration of his new world and the interaction of his body and self with it, only to be met with an unexpected obstacle. Schore reports, "The mother of the eleven- to seventeen-month-old toddler expresses a prohibition on the average of *every nine minutes,* placing numerous demands on the infant for impulse control." (Italics are mine.)

By prohibition, Schore means the mother's NO! or DON'T—and, all too often, physical punishment—concerning some action the toddler undertakes, such as reaching for an object in the grocery store. The impulse control demanded by the mother is selective and arbitrary, determining what is permissible to be learned through exploration and what isn't. While there are times when a mother is genuinely and legitimately concerned for a child's safety and well-being, above all she is concerned that the child learn to mind her and obey her commands as a matter of principle more than practicality. A good child is one who obeys and a good mother is one who has a good child. Both judgments are levied by culture.

The child is impelled by millions of years of genetic encoding to interact on a full sensory level with the events of the living world, through which he builds his structures of world knowledge. The toddler shows great delight in this, and his will, thought, and energy focus with complete absorption and determination on this great building project. It's natural, then, that he resists parental prohibitions every nine minutes, but it is a fact that he is eventually beaten down.

Psychologists refer to the child's instinctive drive for sensory exploration as impulse behavior and insist it must be curbed if the child is to be

---

3. From Allan Schore, *Affect Regulation and the Origin of the Self: The Neurobiology of Emotional Development.* Hillsdale, N.J.: Lawrence Erlbaum Associates, 1994, 30–60.

socialized (civilized) and is to develop a self sense. In turn, the process of breaking down a child's resistance to these restrictions, which is equivalent to breaking his will, constitutes what is conventionally called socializing a child. Of course, as covered in our last chapter, this is not at all socialization, but enculturation.

And here Schore goes into great detail, explaining, "Shame is the essential affect that mediates the socializing function."[4] The authorities Schore quotes assume axiomatically that this "socializing" must be enforced; that prohibiting self-generated impulse actions is absolutely necessary; and that instilling a sense of shame is absolutely essential to such impulse control, leading to proper socialization.

Blake asserts that most self-restraint and restraint of others' actions are sources of much evil. This restraint plays out in our world in a number of ways: Almost all religions are based on restraint of self and others; war is the most extreme instance of restraint of another person; and suicide is the ultimate expression of self-restraint. Inaction and passivity are the ideals of state-religion, while imagination, creativity, and spontaneity are often suspect.

Training a child in how to heed bowel and bladder impulses is carried out in all societies, preliterate to advanced, and in many cases without struggle or trauma. One smart mother, intuitively recognizing the model imperative, simply carried her infant into the bathroom with her each time she had to use the toilet. Within a week or so her newborn spontaneously urinated or defecated in sync with her, making matters of cleanliness quite simple. Fewer diapers!

The same parade of prohibitions goes for the discovery, exploration, and exposure of genitalia, which are always as interesting to toddlers as any other part of their world. In the West, particularly in the United States, where we are still ricocheting from the impact of our Puritan Fathers, these behaviors are seldom handled rationally.

In the final analysis, parental prohibitions extend to virtually all forms of tactile interaction. The untouched child is met with the command DON'T TOUCH! more than any other—and we adults are met with the same

4. Schore, *Affect Regulation and the Origin of the Self,* 200.

words regarding children (to touch a child now carries connotations dark enough that we all must think twice). "Keep out of reach of children"— seen on everything that could potentially harm a child, from cleaning supplies to plastic bags to medicines—has become one of the most common labels in our land, but its message has reached far beyond its original intent. Keeping their natural world out of reach of children seems to be our national passion. In fact, greater numbers of children are brought up in the artificial world of cement, asphalt, plastics, and the virtual reality of television, while fewer each year experience a world of nature and the unfolding of organic life.

THREATENING THE BOND

▼

While every nine minutes a NO! or DON'T blocks the toddler's attempts at world construction, Schore points out that the mother's facial expression conveys her directives as powerfully as language. The mother can accuse and shame a child simply through her look. An accusatory or scolding look becomes a substitute for verbal command and warns the child that his action could break his bond with the mother and bring isolation. This shuts down the child's positive emotional state on which exploration and learning depend, leading to his withdrawal from that exploration out of fear of further threat to the bond with the mother. Schore puts it this way: "The mother utilizes facially expressed stress-inducing shame transactions which engender a psychobiological misattunement."

Even if, in spite of threat, the toddler continues an exploration, such action is then carried out in a negative emotional state and the learning involved will include this negative imprint. "Such bioaffective communications," Schore explains, "trigger an inhibition of the infant states of hyperarousal that support a positive affect." Recall our brief outline of state-specific learning, in which the emotional state present at the time of learning locks in as an integral part of the learning itself.

According to Schore's authorities, the mother's negative prohibitions are necessary to ongoing growth of emotional affect and cognitive knowing, both of which support socialization. Yet Schore describes over many pages how each prohibiting NO! or shaming look brings the shock of threat, interrupts the will to explore and learn, and produces a cascade of negative hormonal-neural reactions in the child. Schore then describes at length the child's depressive state brought about as a result of these episodes of shame stress.

Again, the confusion and depression in the child come from two powerful encoded directives: The first is that the bond with the caregiver must be maintained at all costs. The second is that the world must be explored and knowledge of it must be built at all costs. The mother, with whom the child is compelled to bond, is the major support, mentor, and guide in the toddler's world exploration and learning. When the child, who is also compelled by nature's imperative to explore his world, is threatened by the same caregiver when he does just this, the contradiction is profound. (Laboratory animals can be driven to schizophrenia by conflicting directives, or double binds, in which they lose no matter which way they turn.) The resulting ambiguity drives the first major wedge in the toddler's mind, which, over time, becomes a gaping chasm.

In his research Schore quotes Winnicott, who speaks of the "good-enough mother—[one] who can tolerate inducing stressful socialization transactions in the toddler." Even though she is aware that the child experiences depression, the "good-enough" mother induces shame stress without worrying about it, for she apparently believes, according to Winnicott, that such trauma is "necessary for the child." Winnicott claims that shame stress is necessary for the child to experience if he is to establish a separate identity, and further states that sheltering children from such stressors "is counterproductive for optimal emotional development."

If, however, we examine the emotional health of our populace, as well as our emotional intelligence, we will find less than optimal development, this despite the fact that shame stress is regularly induced in children. Further, this notion of a separate identity at a toddler's young age is, to say the least, questionable. Premature separation of the child from the continuum of life means that isolation, alienation, and estrangement become the foundations on which the rest of the child's emotional life will be based. A separate identity comes in its own good time, as our bonds with ever more expansive realms (family, earth, society,) unfold. And, though the incidence of this is rare, true individuation may come to the child who is allowed to be a child and is nurtured throughout childhood. Forcing a sense of self in the second year of life produces a product of culture that is necessary to culture's maintenance but inimical to nature's intent.

PASSING ON THE SHAME

▼

Use of shame as a socializing technique passes on to the child the very wound inflicted on the parent. As can be seen in the phrase "It's good for you!" which some use as explanation for subjecting children to fear and emotional pain, throughout our lives we act out and then rationalize our shame. Having been shamed, we tend to project our shame on others, looking for shameful acts in them, our judgments always tinged with anger. (Alice Miller addressed this in her classic work on child abuse, *For Your Own Good*.)

Of course, boundaries must be established for the toddler's actions, and caregivers have always provided these from common sense and intuition. Such boundaries set by the intuitive mother are surprisingly few in number, seldom arbitrary, and give a child a sense of security, certainty, and solidity. Throughout time children have accepted such boundaries because they have been programmed by nature to do so. Children want to do the right thing, maintain the bond, win the applause and laurels, as well as avoid saber tooth. Surely over the years, however, these naturally set boundaries and naturally accepted constraints have degenerated, like intuition itself, or have disappeared, along with common sense.[5]

Most of this shaming isn't so much from parents' concern for their child, as rationalized by all of us, but from the parents' own enculturation and serious concern that their own social image might be tarnished by their child's behavior. This personal concern of parents can far outweigh concern for the child's welfare. If their child doesn't conform to cultural expectations, they, the parents, will be criticized, by neighbors, other parents, grandparents, in-laws, the psychiatrist, maybe even the law! This personal

---

5. See Colin Turnbull, Margaret Mead, or Mary Ainsworth and Marcel Geber in their early studies on infancy in Uganda. See Jean Liedloff, *The Continuum Concept* (Reading, Mass.: Addison Wesley Press, 1977) and my own *Magical Child* (New York: E. P. Dutton, 1977) or *Magical Child Matures* (New York: E. P. Dutton, 1985).

fear cloaked by an overtly displayed concern for the child is a major way by which culture perpetuates itself.

As long as the infant is still in arms everyone smiles. How sweet! There are no actions to censure. But the moment the child is up and charging about, everything changes. Censure and prohibitions begin. "Mother's changes are matched by ontogenetic adaptations on part of the infant," Schore relates. *Ontogenetic adaptation* means that nature's inherent genetic plan for development is disrupted, requiring that she must compensate by making new neural connections or rerouting established ones. The *mother's changes* in Schore's statement refer to the rapid shifts in affect of the mother every nine minutes from nurturing to prohibiting once the infant is up and about. Given the means for exploring his or her world, the toddler is blocked by threat of punishment if he does so, which he reads also as threat of abandonment.

Schore points out another ingredient in this mechanism of exploration and prohibition: "Mismatches develop interaction and self-regulatory skills" in the toddler. As used here a *mismatch* means that the toddler's behavior and the mother's expectations clash, which threatens to break the bond between child and mother. Schore's term *self-regulatory skills* is a euphemism for the reactions the toddler is forced to adopt in order to avoid reprisal. These often become a form of lying, a kind of psychological, "streetsmart" ploy adopted by children to accomplish their end.

In doing so, however, as Schore points out, "shame is internalized and becomes the eye of the self looking inward. . . . The other person [the caregiver who originally induced the shame] is then not needed. . . . Shame becomes an imprint, a mental image of a 'misattuned' mother face." Such misattunement between child and caregiver "engenders a rapid brake of arousal and the onset of an inhibitory state." Inhibition is a form of depression; the same hormones are involved. " 'Signal shame' results, an internal mentation alerts the child that [an] external event might be a painful affect." That is, the child develops awareness that an action he is about to take could bring painful emotional reactions.

Signal shame becomes the primary model imperative, blocks the child's natural acceptance of life, and introduces hesitancy and doubt. As the child reaches out to explore, the signal from his internal mentation is: *Stop. You are no good. If you do this, you will be looked at and despised.* If you look carefully at Schore's description of what he assumes the shamed child feels, it's easy to see the empathic rapport he has with the shamed child who is just discovering his initial relationships with his world.

Despite his empathy, that Schore accepts as natural and necessary this childhood tragedy is itself tragic—yet it is a natural conclusion of the enculturated mind. Schore even feels the need to rationalize his empathy itself by claiming, "This imprint [of the shaming caregiver's face] allows the child to regulate his impulsive behavior." So telling here is the word *allows*. The toddler is *allowed* to regulate his own exploratory behavior! What occurs as a result of this entire mechanism is that nature's imperative to explore the world at large is overwhelmed by the greater imperative to avoid the pain of a broken relationship with the life-giving caregiver. What will be developed in the child is a capacity for deception as he tries to maintain some vestige of integrity while outwardly appearing to conform. Living a lie to survive a lying culture, the child forgets the truth of who he really is.

THE WORK OF SHAME

▼

"Shame acts as a major force in shaping the infantile self," Schore points out, but which or what kind of self he is talking about is no small matter. He quotes Darwin: "Shame stress is an essential affective mediator of the socialization process. Shame elicits a greater awareness of the body than any other emotion . . . shaming conditions specifically induce stress reaction." And this stress reaction is lifelong, as evidenced by the current flood of books on shame and its effects on adult life. Rudolf Steiner wisely observes that a child's awareness of body isn't fully formed until about age six, when his consciousness fully "comes down into the body." That is, the child's full awareness of and identification with the body is quite late-forming. Shame breaks into this natural process and the premature awareness that results is a split between self and body, an inner rejection of body rather than an acceptance of self as the whole being nature intended. From this will grow our rejection of the larger body of man and a rejection of the living earth demonstrated in the rape and desecration of our planet.[6]

From citing Darwin, Schore moves onto citing heavyweight Sigmund

Freud, who states that the shift occurring at the end of the toddler period moves the child "from the pleasure principle to the reality principle. And this shift takes place through shame." Note that the toddler is being extricated from the darkness of the pleasure principle and moved to the light of a "reality principle" through the "enlightening" principle of shame! Freud's logic sets "reality" against pleasure in an either-or opposition typical of the dark cultural and religious inhibitions of life. Herein looms the lifelong cultural verdict driving both East and West: Pleasure is bad! Pain is good for you!

Freud outdoes even this perversity by observing that in the late toddler period the child is finally capitulating to pressure and modifying his or her behavior in ways that indicate "deflation of grandiosity and a reluctant departure from primary narcissism." What is being said here about a toddler teetering around in his new world and discovering his relationships with it is this: He is inflated by "grandiosity" and heavy-laden with the sin of narcissism, or self-love. What a revelation to this father of five and grandfather of twelve (most home-birthed and homeschooled or fortunate enough to have Waldorf education)! What we witness is the toddler, a being full of an exuberant inborn love of life and self, becoming the exact opposite, a self-loathing being more suited to live in a Freudian culture and world.[7]

Jean Piaget spoke of a major characteristic of childhood being "an un-

---

6. A segment of the Christian Right has been promoting daily, frequent spanking as a way to ensure that this sinful, "impulsive" nature of our children be curbed. The theory is that until the willful child's will is broken, God's will can't take over. In the diary of the wife of John Wesley, the great Christian missionary to the "heathen" Indians and the founder of Methodism in America, we find an interesting passage in which she describes the pain and anguish caused her by the screams of her children as John gave them their daily thrashing. She tried to console herself with the knowledge that this had to be done, that unless John beat the devil out of them, they would be lost to perdition. This phrase, "to beat the devil from someone," was taken quite literally, arising from the belief of the Protestant mind of the time that the devil resided in those who misbehaved or would not follow orders. Their assertion was that frequent thrashings would make the body of the child so intolerable that the devil would leave it. The same essential notion resulted in the Catholic flagellant movement in the Middle Ages, and is apparently still believed by some.

questioned acceptance of the given." To the young child everything is as it is—wonderful, exciting, inviting, and entrancing—and all of it draws him into an intimate rapport and total involvement and interaction with the world. Once shame is imprinted, however, there will never again be "unquestioned acceptance of the given." Instead there will be a faltering hesitancy as doubt intrudes and clouds his knowledge of self and world. Muktananda considered doubt an evil, as did our great model. Blake summed it up this way: "Were the sun and moon to doubt, they'd immediately go out." Blake knew that this creation brought about by the creator-created dynamic is a huge leap of faith.

The work of shame does not stop with doubt, however. Shame stress brings the same overload of cortisol and depression and withdrawal found in children who experience psychological abandonment or separation anxiety. These pathologies result from loss of, prolonged separation from, or abuse by the caregiver. "The identical . . . pattern observed in attachment-bond interruptions," Schore writes, "[is] brought about by shame stress . . ." Shame stress is a state "characterized by elevated cortisol levels . . . and withdrawal response. . . .

"Increased cortico-steroid levels are also found in twelve-month-old infants undergoing separation stress from the mother," Schore notes, and, "[t]his condition results in avoidance of mutual facial gazing." Mutual face-gazing is the foundation of all audiovisual communication and is primary in all brain development. In some autistic and many depressed children, eye contact, so critical to development in the earliest months, was not available when required and, when offered later, too often indicates hostility. As a result, eye contact is regarded by such children as threatening and is avoided.

Northrop Frye called the accusation of sin the triumph of the death impulse, and Blake pointed out that accusation leads to a complete torpor and paralysis of mind. Schore finds precisely this torpor in the shamed

---

7. Surely Freud has had far more detractors in recent years than champions and I appear to beat a dead horse. Current opinion leaves intact very little of that brilliant man's profoundly wrong life work. While popular conceptions of Freud's thought were hopelessly flawed, his own deep neuroses darkened much of twentieth-century thought and we are still very much under his shadow. As champion of and spokesman for the culture of his time, he is the embodiment of the old adage that nothing persists so tenaciously as a bad idea.

infant and makes a summary statement of our human tragedy when he speaks of the effects of shame and the toddler's "transition from the joyous happy state of effort without stress . . . to a helpless depressed state of distress without effort."

Schore's words should be writ large; they articulate the fall of the human from grace into culture. And how early in the game this fall comes! This surely describes our all-too-real enculturated state of growing depression and distress, a stress arising within us with or without effort, and in adults as well as children.

THE GREAT NEURAL PRUNING

▼

This brings us to the most critical of all Schore's observations from his twelve years of work and 2,300 research citations. Delving into the negative aspects of our biology, this observation is the pivotal point of part 2 of this book. But first a reminder: The prefrontal lobes are experience-dependent; the environment must furnish the appropriate stimuli if full growth is to take place.

Note, then, that the prefrontals form their major, large-scale synaptic connections with the emotional-cognitive brain in the first year of life because that is the period when most nurturing takes place. In the final weeks of that first year of life, at about the eleventh month, a superabundance of dendritic links are grown between the prefrontals and the cyngulate gyrus, the foremost part of the emotional brain, in that critical area called the orbitofrontal loop (see chapter 2). Nature makes these neural connections in excess, Schore notes, a fail-safe overproduction. This prepares for the general brain growth spurt that precedes the beginning of the toddler's walking, talking, and critical exploration of the world and his relationships to it.

Yet, shortly after that major preparatory growth spurt in the prefrontal-limbic connection, nature deconstructs those very neural structures—and thus the very orbito-frontal loop that she has just established! Recall that the prefrontals are nature's latest neural creation, and this orbito-frontal connection is the fourth brain's link with the ancient emotional-cognitive brain and, through it, with our heart.

Schore relates that the emotional shaming experience the toddler undergoes brings about a "degeneration and disorganization of earlier imprinted limbic circuit patterns . . . [and] produces a rewiring of orbitofrontal columns."[8] He then details not only how the actual neural growth of structure and hormonal balance in the child are impeded by shame, but also how shame actually brings about the deactivation, severance, and pruning of those very superabundant connections that have just been established between limbic and prefrontal systems. In Schore's words, "a period of maximum synaptic excession occurs within the human prefrontal cortex at the end of the first year and thereafter declines. . . . Such alterations are known to be related to functional use-disuse."

The worst is yet to come, however. Far more devastating than this pruning is that nature then brings about a corresponding increase of the connecting links of the emotional circuits in this cyngulate gyrus with the lower survival fight-or-flight structures of the amygdala, that neural module linked directly with our ancient defense and survival system in the reptilian brain. In this way, a sharp curtailment of connections with the higher, transcendent frequencies of mind and heart is brought about in order to shift growth toward the lower, protective survival systems.[9]

This is, again, just what we observed happening to the brain of the infant in utero when the mother is subjected to anxiety. Nature has again provided an excessive amount of neural material for a movement toward higher intelligence, and again has had to retreat on behalf of survival. This will happen again and again, particularly in the parallel adolescent period when corresponding growth spurts once more take place between the emotional brain and prefrontal lobes. (Occurring at adolescence is an advanced form of maturing analogous to that of the early toddler stage, when emotional connections are again uppermost in importance.)

Schore rationalizes that this might be a way of pruning neural structures that are needed in an early, temporary stage of development but are not needed in the building of a "social self system." This notion of temporary neural structures was the explanation long given for the extraordinary neural pruning nature employs right before birth. We now know this prenatal pruning results from nature's excess production of neural cells to cover both

---

8. Schore, *Affect Regulation and the Origin of the Self,* 252.

9. "Shame stress deactivates the ventral tegmental and activates the lateral tegmental limbic circuits . . . bringing lowered opioid levels . . . and reduced growth of the sympathetic excitatory mesocortical dopamine systems."

ends of the spectrum in infant brain growth in utero: either a move for higher intelligence or a shift to the lower defense system This either-or situation is resolved, as usual, by the environment. A toddler "pruning" that takes place in the weeks immediately following the very growth spurt provided by nature in that neural group follows the very same pattern found in utero and in the months following birth. It is as though our life's intelligence puts out feelers to test the climate and sends ambassadors to see if negotiations can be established, but finds it must retreat to defense time and again.

Schore refers to "activity-dependent preservation of synapses" making up the neural fields of brain. That is, *use it or lose it* is nature's dictum, which rather balances her largesse. There is a precise devolutionary process occurring here. At this most critical time, when the toddler begins exploring the world, the prefrontals lose the very synaptic connections they have just made with the limbic system and, through it, with the heart, the connections prepared for during the in-arms period and throughout the general nurturing period of that first year. When all the rest of the brain is growing at its greatest rate and enormous world exploration is supposed to take place, the prefrontal-emotional connection is cut back, withdrawn. Which area of the brain is instead receiving that energy, attention, and stimulus for growth? Of course, it is the hindbrain and its emotional loop, busily building defenses against a world that betrays and can't be trusted.

At the time of the toddler's brain growth spurt and again at the adolescent's, nature asks, "Can we go for higher intelligence now, or must we defend ourselves again?" Our toddler's actions clearly display this ancient battle between devolution and evolution playing out once more. "Dyadic shame-regulating transactions in [the toddler] generate permanent effects in the fronto limbic cortex [orbito-frontal loop]," Schore observes, and indeed they do.

This loss of prefrontal material is brought about because as the caregiver becomes the "socializing" parent, emotional deprivation takes the place of nurturing in that second year—and the excited, exuberant child is turned into a "terrible two." More is involved here than use it or lose it—we witness a major shift from higher levels of intelligence to lower levels of defensive instinct, a natural survival reaction the child's system must make to a harsh emotional environment. And we applaud this as successful "socialization" of a child.

NO TIME FOR NURTURING

▼

In his work, Schore pleads that each shame-inducing episode be followed immediately by sufficient nurturing from the caregiver. Nurturing immediately after prohibiting or shaming, Schore points out, "reestablishes the bond, which action not only alleviates, counteracts the negative effects, but brings about a positive learning. " Although this again smacks of rationalization, making a virtue out of necessity, Schore gets things back on track toward what's truly best for the child: "Nurturing the infant . . . induces long-lasting changes . . . in the adult frontal cortex . . . and permanent modification of later [hormone production] which would . . . increase exploratory behavior and emotional response [and play] an important role in regulating higher-order information processing." [10]

Though there is acknowledgment of the necessity of nurturing to counteract the negative effects of both shame stress and prohibition of exploratory behavior, with a negative prohibition occurring every nine minutes and the withdrawal depression in the child generally lasting much longer than that, which parent can or is willing to take the tremendous time needed to nurture the child sufficiently? And how many actually do? Why not, instead, work to eliminate this whole cultural travesty by exposing the unquestioned acceptance of negative prohibition and shame as tools for behavior modification? We must show the folly in the assumption that "socialization" at this early stage of development is necessary at all, much less beneficial. Make nurturing, care, love, and a buoyant, happy child the entire criteria of social success in parenting. Let parents be known "by their fruits" that give later peace, not violence.

Recall Patricia Goldman Rakic's statement: "The ultimate function of the neurons in the prefrontal cortex is to excite or inhibit activity in other parts of the brain." In prohibition and shame we excite the most destructive systems and inhibit the creative ones. As of this writing, failure of nurturing has led to the rising inability of our young people to modify primitive impulses and behaviors. By 1995 an average of eighteen children per day in the United States were being struck by bullets from other children's guns. Some six thousand a year die from those wounds and violence has become

---

10. Schore, *Affect Regulation and the Origin of the Self,* 244.

virtually a national security matter. We can't seem to build prisons fast enough and thirteen- and fourteen-year-old children—both boys and girls—are incarcerated in penitentiaries. Sixteen-year-old girls have delivered babies in prison under horrifying conditions. To say the least, our "socializing" tactics are working poorly for our wounded society. (See figure 9.)

To the epidemic of children shooting children add the increase in child suicide. Up to the post–World War II period, no suicide under the age of fourteen had ever been recorded. While this may have been due in part to lack of reporting, a lack of acknowledgment, or both, even with full reporting the statistics then could not approach the current rising child suicide rate in the United States. In 1991 it extended to children as young as three, with an attempt occurring every 78 seconds. Some six children a day succeeded (which points to the facts that we have an excellent 911 emergency system and that suicide proves far harder to bring about than most children are aware of). As of the year 2000, suicide has become the third highest cause of all deaths in children between the ages of five and seventeen. Far

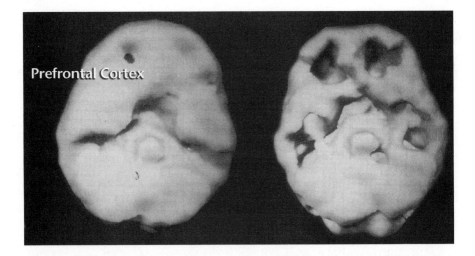

Figure 9. Thanks to the latest brain scan devices, here are the top views of the neocortex and the prefrontal lobes of two living people. On the left is the scan of a normal, nonviolent person; on the right is the scan of a violent person. Note that the neocortex (the lower portion of the scan) appears fairly uniform in the two, while the prefrontals show dramatic differences. Such scans provide clear evidence of the biological damage afflicting violent and/or criminal people and point out the critical handicap they face as a result of this damage.

more suicides are attempted by girls than boys, yet boys far outnumber girls in actually completing the act. There is no historical precedent for this phenomenon at all, and though teachers are very much aware of it and do not ignore it, it is largely ignored on any national or institutional level. While we often hear the constant, shrill commands of NO! and DON'T at every hand, from cradle to grave, seldom do we see nurturing and love. The price of the toddler's arbitrary compliance with our own shameful action is paid over and over, year after year, by our whole nation.

THE GUILT LIVES ON: THE LIFELONG SEARCH FOR ABSOLUTION

▼

The induction of shame is a blatant form of the accusation of sin, and because most of us have heard this and been the recipients of such accusations from the beginning of life, we unconsciously and impulsively inflict the same on our children. Schore's quote about this shame perfectly articulates the tone of the accusation: "You are no good. Your action is bad." Shamed in this sense, we forget who we are. We actually become the protective mask we adopt to shield us from the accusing fingers pointed toward us. Cut off from our spirit, we spend the rest of our life trying to prove our innocence.

And this brings us back to the chapter's beginning—to our roof-brain chatter or compulsive stream of consciousness. It is from our state of shame that this inner speech arises, bubbling up without cessation, full of accusation and fault-finding as it attempts to cast out of us the dark shadow of shame forced upon us from infancy.

That all of our responses to shame are replicated and significantly intensified during adolescence is a major subject of discussion in itself. Suffice it to say that with the development of a new body, the extensive brain growth spurts, and the onslaught of sexuality with its attendant massive shame-inducing restraints, this time in life provides multiple opportunities for the sentence of guilt to be pronounced again and again on us, making us fit subjects of culture, fit vehicles for violence, and appropriate consumers of massive cultural counterfeits.

As we grow to adulthood, the good news proclaimed long ago—that we are made in the image of God—is rejected, the cultural self accepted, and we spend our life trying to expose the lie against us for what it is. Many of us are driven to try to prove our worth (or innocence) to our accusing world by earning success within it—and at any cost. While we work at this, culture seductively holds up to us models of every description for achieving these desperate ends, and we, the accused, become what we behold. If we have the mettle, we stop at nothing to "get to the top" and so be declared clean by the high priests who hand out the cultural laureates. Each accused soul scrabbles for gain at any price because wealth alone is deemed proof of authenticity and freedom from censure. Always gained at the price of our fellow humans and now at the expense of our living earth, these seductions of spirit and energy spin our culture along its path.

Thus it is that our innate passion for transcendence is sidetracked and derailed, ironically empowering the cultural morass we see at every hand.

But many of us are not only incapable of finding absolution but seem to backslide after every step as well, succumbing to what culture would have us believe is our predetermined behavior. Among the greatest of cultural lies is that we are, by nature, violent, and that only chaos would prevail without the order imposed by law and the shame stress that powers it. We automatically serve culture and perpetuate its violence simply by believing this lie—yet it proves to be self-fulfilling and prophetic. Here in America, two million of our brothers are in prison. Often our minorities seem selected for cultural stoning, scapegoat victims and captives of our lie. Just as enemies abroad must be created to keep the making of munitions alive and well, we force into ghettos, minimum-wage jobs (at best), crumbling schools, and minimal health care an entire class of people, subtly accusing them of moral and ethical failure to live up to our lofty standards. We thereby create a stream of the lawless and criminal that we point to as proof of the need for ever more law and repressive order. Like the Romans at the Coliseum, we keep our masses entertained, our populace entranced, as our oppressed minorities supply our entertainment industry with ever-fresh materials and scripts for its obscenities.

If that "higher self" within us is to be born, or better, revealed, it is our cultural self that must, in effect, die. If we are to follow the example of our great model, it is this world of culture that we must spurn uncompromisingly while, at the same time, we love and forgive ourselves and each of those caught with us in this maze.

The third highest cause of child death would not be suicide were the example set by society the one emerging from the heart and the good news planted in it two millennia ago. Corporate heads or leaders of state who rape the planet and its peoples would not manifest were the heart and gospel our model. As long as culture is our model, we will get more of what we have. How much longer, we might ask, can enforced cohesion stave off chaos?

In the next chapter, the most negative of this negative part 2, we will go head-to-head with one of the earliest and strongest objects of my love: the church.

▼

EIGHT

THE GREAT ACCUSATION

State religion produced its master piece . . . in the god of official
Christianity . . . invented as a homeopathic remedy
for the viral teachings of Jesus.

—Northrop Frye

Civilizations that experienced no warfare—and so, possibly, no violence—may have existed some ten thousand or so years ago. Consider the
Harapi, an apparently advanced and civilized people whose cities and towns
stretched from the Ural Mountains to present-day India long before the
Egyptian and Middle East civilizations appeared. Excavations show that
these orderly communities used common dimensions and weights and brick
of the same dimensions and laid out cities on the same symmetrical ground
plan. In addition, they all had running water, underground sewage, and a
form of common food storage. And as far as can be determined, they had
no weaponry of any kind, and throughout the whole vast complex there are
no signs of warfare having occurred. I am told that recent excavations in
China have unearthed statuary, apparently of ancient sages, that depict individuals with extraordinarily pronounced prefrontal lobes, a phenomenon
that probably can take place only in a prolonged era of peace and tranquil-
lity (if you recall from chapters 2, 5, 6, and 7).

Evidence is far stronger, however, that our species has struggled in a sea
of its own blood and carnage for several thousand years now. As we have
seen, an evolutionary advance of mind and spirit can be lost as our species
again retreats into its survival mode—sufficient hindbrain bought at the

price of the forebrain. But evolution is always on the prowl, looking for the opportunity to shift us into a higher mode of functioning.

We speak of the Golden Age of Greece, which lasted but a brief flicker of time (half a century or so at its height): Its achievements resulted from the pastimes of a minuscule number of enlightened people who had the leisure for such things—and from the massive number of slaves who enabled the existence of such leisure. Similarly, the glory that once was Rome rested on a constant bloodbath of violence, pillage, and slavery that has few rivals in history. Even before Rome's entry into that arena, the Middle East was a constant ferment of strife, incredible cruelty, and violence inflicted by humans on other humans, as the Old Testament, archaeology, and records attest. The constant struggle of humankind seems to have been to survive itself, the pillage and ferocity its own members levied against each other. The paired cause and result of this phenomenon I have summed up as *culture*, the real substrate of all the varied and occasionally colorful episodes that make up our history.

Into the ugly cultural scene of two millennia ago there was injected a minor, scarcely noticeable event in the Middle East: the Crucifixion, traditionally marked with a capital *C* because, for centuries, the ubiquitous Romans crucified en masse. There was, however, an evolutionary impetus behind this particular Crucifixion, and though that impetus fell victim to the very cultural effect it sought to break through, a warp in culture's history took place as a result. Gil Bailie rightly claims that this historical warp was due to the Crucifixion bringing to human consciousness the dawning awareness of the plight of the victims of our carnage rather than just the glory of the victors and their spoils, which cultural history always emphasizes. The dawning impact of that awareness of the victim has been ruefully slow in taking hold, however, because of the counterinfluence of culture, and may be only now, in our own time, appearing on any significant scale.

This chapter explores the way in which culture warped the breakthrough wrought by the Crucifixion, resulting in the creation of Christianity as another powerful form of culture itself, producing its own centuries of victims and dampening further the painfully slow and paltry influence the Crucifixion has had on our violence. Thus, while nothing since that historical event took place has been the same, no fundamental change seems to have taken place either. We simply cloak under different political, economic, social, and religious terminology and rationale our current versions of carnage wrought upon each other.

For two millennia now we have witnessed the strange paradox cited at the opening of this book: a constant parade of lofty ideals negated by abominable behaviors—the deadly struggle between transcendence and violence. Two thousand years of weeping, wailing, hymn singing, and praying led us not to the kingdom of heaven but to such hallmarks of hell as Dachau and Buchenwald, Hiroshima and Nagasaki, the bombing of Dresden, the rape of the planet, millions of abandoned children—the list is endless and is always being updated. Though primary perpetrators have been Judaic-Christian nations, few religious persuasions are without stain. The force behind such violence: culture.

As the cultural counterfeit of transcendence, state-religion is the equivalent of the underground run by the establishment in George Orwell's novel *1984*. Through our longing for transcendence we are enticed by the religious counterfeit, which leads us, unbeknownst, back into the boundaries and bondage of culture, though filled with a sense of righteousness and virtue.

We vaccinate against a disease by taking a dead virus and injecting it into our body so that our immune system will build antibodies to counteract the threat and when the real thing comes along will throw out the rascal without our knowing a confrontation took place. Religious upbringing operates similarly; it often injects the notion of a dead god into our psyche and when the real transcendent force arises within us throws it off without our even knowing that an encounter with the real took place.

The word *satan* means the "the accuser." Keeping this in mind, consider that Blake perceived state-religion, or church, as allied with Satan—church, after all, accuses us of sin. (Admittedly, this becomes a mutual finger-pointing typical of the cultural hall of mirrors we live in.) Blake, for his part, allied himself with the devil, who, in Blake's cosmology, was Satan's adversary. It is with Blake and his view of the church that I take my devilish stance, pointing my finger as well. My claim is that accusation, a dark addendum to the gospel spread by the church, nullifies the light of the gospel given to us by that figure on the cross. You can't have both the darkness of accusation and the light of the gospel—the darkness dispels the light.

Recall the devastating effect that deadly NO! and the accusation of shame or sin has on the toddler. Precisely the same effect occurred to all on hearing this dark side of the news spread by the church (which speaks to Blake's assertion that the accusation of sin brings a torpor and paralysis of mind). Christianity fosters a father-child identity for the relationship of God and man, at least in theory. But through Christianity, with what kind of God are we identified? World history would have been dramatically different had that loving Father of Jesus been the message spread by the gospel, but such was not the case.

A reader might wonder why we dig up such ancient history, which may seem of no consequence today. But nearly every negative we suffer today spins directly from that history. The effects of enculturation analyzed by Allan Schore have their genesis in the cultural events of two millennia ago. The Christian institution has been and still is the mainstay of our Western culture, and we live and breathe the results of this now as before and will continue to do so as long as culture stands. Our beliefs or nonbeliefs have no more to do with culture as process than they might with gravity. But our beliefs are reflected in the dynamic of creator-created and we live them out without being aware of it.

Through the use of myth and superstition engendered by fear, the religious institution has woven a rich tapestry of betrayal around the gospel, a fabrication I hold as typical of culture itself. In exposing this myth and make-believe, some glimpse of the original good news may still be found, fresh and new from being obscured for so long. The gospel was a cosmology, a description of the creator-created dynamic, and it was, as the name implies, truly good news.

CREATING THE MYTHICAL STORY OF JESUS

▼

The word *evangelist* means "one who spreads good news." The problem for the original evangelist was that the good news of Jesus was at a radical discontinuity with the mind-set of the culture of that time, just as it automatically is to ours today. The early evangelists didn't hesitate to correct this shortcoming in the gospel by changing the radical nature of Jesus' message itself in order that it might be heard and accepted, thereby becoming the new culture. In interpreting the gospel so that it might be heard by the old mind-set, they put the new wine in the old wineskin and no one was the wiser, as we shall see. That which was thought to be a new mind-set had become the old in new form, with all the old murderous issues still intact.

Far more serious than their initial uncertainty as to how to get across their interpretations of the Jesus event was the tendency the evangelists had to demonize those opposing them. Elaine Pagel's brilliant study *The Origins of Satan* shows the polarization of the Jewish communities brought about by this negative evangelism, and by the evangelists' constant rewriting or retelling of the gospel itself. Through the argument over interpretation, as believers struggled in the scrabble to establish a church or organized body of followers, the good news of the God-human relationship disappeared, bringing ever more extreme translations of the original event.

Among many techniques used for translation, the evangelists adopted a series of mythological overlays for Jesus and the events of his time on earth, a myth that gave new life to the ancient practice of a father sacrificing his first-born son to appease the various gods and goddesses of the ancient Middle East. The story of Abraham in the Old Testament's Book of Genesis relates the Hebraic abandonment of this sacrificial custom, sanctifying a substitute that strengthened Jewish culture and led to the history of the temple, where sacrifices were held, as the cultural hub. But the rebirth of this powerful archetype of sacrifice in its new gospel dress put Christianity on the map, and the principal revivalist of this association was Paul.

He and the evangelists simply reversed the logic of the ancient practice: A wrathful god sent his own son to earth and in effect sacrificed his son to his own anger, a move made, strangely, on behalf of the erring race who had offended this moral governor in the first place. Making sense of a god who fathered and then sacrificed his son to his own wrath took generations of overlay to fill out, with half a century elapsing before the first written accounts of this mythical hybrid story appeared and set the stage for all future accounts.

This process of creating the mythological life of Jesus, including the very reason for his life, fits Mircea Eliade's pattern—recall Eliade's observation that great myths are overlaid only on great people. Jesus was retroactively fitted out with such a background through a long, organic process of imaginative growth that allowed many storytellers and chroniclers to add their imaginative pieces until the final distillation called the New Testament was hammered out. In the long and often bloody turmoil over whose mythical interpretations of Jesus would be accepted, culture was strengthened. Its new dress of Christendom became its new armor, and the gospel all but disappeared.

The Christian movement didn't really gain momentum until the archetype of father-son sacrifice was wedded with the invention of the Second Coming, or imminent return from the heavens of Jesus as the Christ. This hypothetical end, always just around the corner, even today, justified whatever means could be found to alert the populace to prepare for that soon-to-come demise of the whole world. Jesus' actual "return" was as the Holy Spirit, or *pneuma*, which breathed as tongues of fire on the remaining disciples at Pentecost. This extraordinary group epiphany should have lit an inextinguishable fire in those men, a fire that should have in turn lit the world with a new vision of humankind and its potentials had those disciples not suffered a failure of nerve.

The more the evangelists used this pitch of the coming Judgment of a sinful species and divine retribution by a wrathful god, the more convincing it became. And once implanted in the species psyche, as with any powerful negative, it was difficult to erase—a heavy negative that is resonant with ancient archetypal images can trigger our primitive survival reflexes and strategies through connection to our defensive old brain. Recall that once these survival patterns are established, they are not very negotiable. When the gospel, the good news that should have nurtured and given strength, instead threatens and condemns, the light goes out for us much as it does for the exploring toddler whose nurturing caregiver suddenly becomes the harsh judge of enculturation. When the good news of our unbreakable relationship with our creator was saddled with the threat of coming judgment and condemnation, the threat crowded the God of love right off the stage. Powerless love doesn't sell; guilt and sin do.

Twenty centuries of this intriguing theology of sin, guilt, and damnation, with the hope of possible redemption if we jump through all the hoops just right, served to create within humans a deep sense of the validity of that very sin, guilt, and hovering damnation. It is the one injected archetype from which we haven't recovered, even in this age of science, just as some of us have not recovered from the shame induced in infancy, which we carry throughout our lives and which colors every event. Remember that the original cause of anxiety is dismissible once fear has been induced. Anxiety persists by filling its empty space, once filled by its cause, with the ever-changing content life brings forth, turning the new into a variation of the old. Because Jesus broke just about every law of the authorities of his time, we might examine our own concept of love versus law and authority. The difference is precisely that between hindbrain and forebrain.

Through his mythological overlay, Jesus has become, as the post-Hegelian Ludwig Feuerbach hypothesized, the most powerful figure in history for projection of all our highest aspirations and our most sublime ideals, resulting in a field effect of tremendous and ever-growing power—which is exactly as it should be—but the shadow of the never-ending cycle of guilt, sin, and redemption has acted as an equally powerful counterforce, essentially nullifying much of the potential of the positive field effect that might have lifted humankind above its violent bonds.

OLD GOD IN NEW VESTMENTS

▼

In addition to the revitalized (if inverted and reinterpreted) sacrificial myth, the evangelists, following Paul's lead, established a relationship between Moses and Jesus—between law and love. They paired the ancient Hebraic testimonial of God and human with the newest, in order to make their product acceptable and desirable to the old temple customers. Thus the crack in the cultural egg, here represented by the cross, sealed quickly and became a cultural support, strengthening culture's protective shell. Because the new story was a variation of the old, the loving father, for whose entry into history and consciousness Jesus lived and died, was converted back into the god of Moses, the real backbone and sinew of both the Second Coming and the revised standard version of the gospel, a version that sold well (while the original may not have).

In and of itself the Old Testament was a magnificent historical legend of a remarkable people's evolution of and growing enlightenment on the nature of God and human. In its earlier era, this evolution centered on the fiery figure of Yahweh, a character as paradoxical and contradictory as the long-evolving awareness of Israel itself. The evangelists' adoption of this thunderous creator of violent jealousies, judgments, and vengeance represented a theological throwback of a thousand years. But in tying New Testament to Old, the evangelists gave new life to this fiery and fitful God who easily replaced Jesus' Father, the giver of good and perfect gifts, in much the way that a judging Christ replaced that forgiving Jesus.

This resurrected New Testament tyrant, dissatisfied with meting out plagues and pestilence for misdemeanors, as he did in the first accounts about him, instead meted out eternal hellish punishment for sins. And, just to turn the screw a bit tighter, these sins may have been committed before we were born, the nature of which people might not even grasp! We were damned by our very conception unless there was divine intervention through Christ—so admit your sinful nature quickly and buy into the system while there is still time!

This sales pitch, still going strong today in fundamentalist, evangelical, and some Catholic quarters, puts Madison Avenue to shame. One institution does it all: induces the illness, diagnoses the disease that results, and sells the victim the antidote. (This, ironically, is the very same process employed by television, the whole commercial world, and the global economy through the use of what Gil Bailie calls "mimetic desire," our enculturated compulsion to do as others do and have what they have.)

PAUL'S INTELLECT AS CULTURAL BACKLASH

▼

The emerging organization of evangelists and believers called church or "body of Christ" had as its backbone the evangelist and apostle Paul. His experience of enlightenment on the road to Damascus was no doubt genuine enough, though his interpretation of it is a real question. Paul seems to have brooded over the interpretation of his revelation for several years, as did Jacob Boehme and others who have had revelations. Finally he surfaced actively among the evangelists to try out his aforementioned saga of the atonement of human and god through ritual sacrifice. This invention bore but the faintest similarity to the original Jesus event, but Paul's genius, like Freud's in our own day, should never be underestimated, nor his lengthened shadow discounted, for that shadow's darkness still covers us.

At any rate, Paul's endless and legalistic interpretations of the bits and pieces of the gospel left almost nothing of the original but acted as a catalyst to bring into a more or less cohesive whole those often conflicting fragments. Even the Paraclete, the Holy Spirit or intelligence of the heart that Jesus introduced and modeled, and upon which his gospel depended, suffered Paul's editing—in this case being deleted altogether. Paul seemed to know, on the surface, something of the conflict of love and law and is considered by some scholars to have created Christianity out of his scorn for and rejection of the legal system in which he had played a part as well as for Jewish temple practice and life.

Though Paul knew a great deal of the law's workings and language, he hadn't a clue about law as a cultural force and gives no indication of grasping the fact that Jesus' way is antithetical to the concept of law in its totality. It was primarily through Paul that Jesus' way was converted into the cultural effect from which Jesus sought to free us. Paul represents the roaring return of intellect to replace Jesus' intelligence of the heart, and he did this with a thoroughness that is astonishing. He was one of those brilliant intellects and systems builders who must tinker with every issue or event and erect great edifices of thought and invention around it, often obscuring the thought or event itself. Thus Christianity became the lengthened shadow of Paul, not Jesus. And though his intellect translated as a bewildering, convoluted logic, he set the stage for the two millennia of equally bewildering theology that followed.

Primary examples of the absurd contradictions in Paul's intellect can be found in his passages about love, some of which, ironically, are as beautiful and familiar as the sonnets of Shakespeare. The opening passages of Paul's letter to the Romans reveal the foundations of his invention, a christology that bears almost no resemblance to Jesus' way but nonetheless forecasts the history of Christendom itself. He writes to his Roman recruits of "divine retribution revealed from heaven and falling upon the godless wickedness of men . . . the day of retribution when God's just judgment will be revealed, and he will pay every man for what he has done. . . ." Paul's God, who gets even with everybody through a retribution that is wholesale and devastating, is not quite the Father who gives good and perfect gifts and judges no man, but his God proved far more popular. Culture is based on fear and loves its own. An enculturated mind is culture itself, a bundle of anxiety in a dying animal (with apologies to Yeats).

Paul gets into real contortions of logic and law when he addresses the Mosaic Law that revealed the true face of the christ with which he replaced Jesus: " . . . [T]hose who have sinned outside the pale of the Law of Moses will perish outside its pale and all who have sinned under the law will be judged by the law . . . *on the day when God judges the secret of human hearts through Christ Jesus."* Don't read too quickly that last italicized phrase. The fall of man is repeated once again within it. Not only has a god of love and forgiveness disappeared, but Christ Jesus now becomes God's instrument for the dirty work of judgment itself. Where now is nonjudgment and forgiveness?

The following quote from Romans summarizes what happened to Jesus' gospel as clearly as any: "[D]o not seek revenge, but leave a place for divine retribution." Apparent here is the subtlety of an archaic theology in a sly play on and appeal to our undercurrent of rage, promising us, in effect, that vengeance will be ours in the divine retribution of the new Christian order. The Christians have been playing this one out for two millennia. Gil Bailie and René Girard show clearly how this desire for vengeance keeps our violent culture spinning in its cycles. Forgiveness disappears at this point, replaced by the smug gloating that our enemies will be dealt with by the invisible stick we carry: God's coming judgment. "Justice is mine says the Lord," Paul quotes, "I will repay." The price of this repayment is the gospel of love, whose loss is incidental to the gain of sacred vengeance Christians have wrought for two thousand grisly years.

Having revived Old Testament notions of justice, the plot thickens with Paul's pontification on judgment. Consider carefully this excerpt from 1 Corinthians 6, announced to his new fellowship of believers: "You are judges within the fellowship. Root out the evildoers from your community . . ." This call for exorcism is followed by his far more generic and universal proclamation that will ring down through the ages: "It is God's people who are to judge the world." (This is a chillingly dangerous viewpoint and suggestion that foreshadows such logic as that of the Holocaust and other ethnic cleansing, particularly when it is held as the verbatim word of God.)

Paul writes this concerning his Roman recruits who take too seriously Jesus' injunction that love is above law: "Every person must submit to the supreme authorities. . . . There is no authority but by an act of God, and the existing authorities are instituted by him; consequently anyone who rebels against authority is resisting a divine institution." Paul speaks here of Roman authority, which rested on Roman law backed by the army, though his observation was made equally to strengthen his own flanks in the scramble for church leadership and would later be applied to church authority itself, with its divinely sanctioned armies. Paul tells us that cultural authority is authorized by God, a concept compatible with Moses, perhaps, but the complete antithesis of Jesus' way.

His thoughts concerning authority are continued in his discussion of government, which, he writes, is "a terror to crime [but] has no terrors for good behavior . . . [for governments are] God's agents, working for your good . . . [therefore] discharge your obligations to all men, pay tax and toll, reverence and respect to those whom they are due." It isn't just that the grounds are laid here for such chicanery sixteen centuries hence as the divine right of kings, and echoed, after another four hundred years, by a host of flag-waving evangelists in that darker side of American politics. More seriously, an unbridgeable gulf lies between this statement and Jesus' observations about love making law obsolete.

In no way would Paul grasp the subtlety of that famous statement of Jesus concerning paying tribute to Caesar. We give Caesar that which is his in order to be free to give to God that which is God's—our heart, soul, and life itself. Consider the delicate stance of agreeing quickly with the adversary lest he deliver you to the judge and prison, which is resonant with the saying of the Sufis that only a fool is honest with the dishonest.

Consider, too, the equally revealing nature of the comment Jesus made to the man picking corn and eating it on the Sabbath. The fact that it wasn't the man's corn to pick and eat was but half, and the lesser half, of the issue. Breaking the Sabbath law by picking the corn was the greater issue. So Jesus said to him: "Man, if you know what you are doing, you are blest. If you know not what you are doing, you are accursed and a common criminal." The intricate and delicate subtleties of functioning from the intelligence of the heart are simply not comprehended by an intellect such as Paul's, which is involved in tortuous, endless arguments that can never be won—and that everyone only loses.

STANDING IN THE GATE

▼

At any rate, Paul couldn't see what Jesus saw—that there was no difference between Roman law and temple law except that temple law "stood at the gate and didn't let anyone through," which drew Jesus' ire. Jesus was, of course, speaking of temple law standing at the gate to the kingdom of heaven within us, which translates as intellect blocking the intelligence of the heart. As to the two sets of law, the Roman, enforced by vast armies, robbed you of money and materials; the other, Mosaic Law, enforced by temple authority, robbed you of your soul. Of the two, the latter was the more fatal, according to Jesus. No great subtlety there. Jesus would point out the hypocrisy and viciousness of both positions, however, and we should note that he did not take a stance against or for either.[1] He stood for the crack in the egg, the narrow opening found in the law of the excluded middle of our logic.

Women figured prominently on Jesus' scene and in his consideration of them he stood squarely—dangerously—opposite the cultural practices of his time. We have only to note his forgiveness of the woman at the well; his intercession on behalf of the adulteress about to be legally stoned by the mob; his defense of women against the gross injustices of the divorce laws of his time; and his willingness to eat and consort with women of ill repute—all subversive acts in that cultural climate. Some gnostic gospels give women a very high place in Jesus' hierarchy, even considering the "beloved disciple" of John's Gospel, that quiet, shadowy presence at virtually every scene with Jesus, to have been Mary Magdalene. James Carse picked up on this in that strange, surrealistic little gem of his, *The Gospel of the Beloved Disciple.*

But Paul put women back in their place, and quickly reinstated patriarchy and the authority of a new temple priesthood. On women he is no less assertive, cocksure, and reactionary than in regard to law, justice, and government. In 1 Corinthians 6 he says: "It is a good thing for a man to have nothing to do with women . . .," though he qualifies this with his admission that it is "better to marry than to burn" even as he admonishes his followers to be strong in will as he is and resist such weaknesses. Lest a man let his need for such love get the upper hand, Paul keeps the scales weighted his way: "While every man has Christ for his head, women's head is man."

He doesn't hesitate to make ample use of shame as his new enculturation device, as in Romans 10: "Everyone who has faith in him [Christ] will be saved from shame . . ." But more to our focus here is this insight given in 1 Corinthians 6: "A woman brings shame on her head if she prays or

---

1. Elaine Pagels delineates with the greatest clarity and simplicity how the few scant original sayings of Jesus were reshaped time and again out of expediency, to gain converts and defend different political positions and gospel interpretations. That the Hebrews themselves became the scapegoat and target as the number of gentile converts grew was a slow but deadly turn of events. One could say the seeds of Holocaust were sown long ago.

prophesies bareheaded." His reasoning behind this strangely Islamic pronouncement, leading to women being barred from a church if bareheaded, is even more revealing: "A man has no need to cover his head because man is the image of God and the mirror of his glory, whereas woman reflects the glory of man." Paul repeats much the same litany in Ephesians 5 and echoes it again in Colossians 3, clearly outlining the supremacy of the male and the inferior status of the female. In Ephesians he urges, "Slaves, obey your earthly masters with fear and trembling," as he likewise admonishes women to obey their husbands.

In his first letter to Timothy, Paul explains: "A woman must be a learner, listening quietly and with due submission. *I do not permit* a woman to be a teacher . . ." (Please note the declarative tense of this astonishing prohibition, which I italicized lest its importance be overlooked.) He continues: " [N]or must woman domineer over men. She should be quiet for it was the woman who, yielding to temptation, fell into sin . . ."—and of course dragged down poor innocent Adam with her. Thus Paul positions his archetypal Eve—and therefore all women—as the originators of original sin, that dark workhorse Augustine rode to the heights of sainthood and of which there is not a whit of suggestion in any of Jesus' own words or actions. Following Paul, this misogynist virus wormed its way into most Christian doctrine and the many versions of the gospel that followed. Very little of the New Testament or gnostic writings escapes this Pauline inversion of Jesus' way. In addition, in Paul's revival of Eve as every woman, he gives himself grounds to bar women from holding church positions within his own jurisdiction, an exclusion picked up by all of Christendom and holding for close to two millennia.

"I do not permit women . . ." In this statement Paul doesn't even bother with the usual guise of divine sanction through the preface "God decreed that women . . .," or "the Lord said that women . . ." Instead this is his flat-out imperial decree: *"I do not permit . . ."* He has become here his own divine sanction, and his word—concrete, tangible, largely parable- and metaphor-free, legalistic, heavy handed, and crystal clear—became a principal substrate of the revised gospel and the formation of a New Testament. Paul appeared on the scene only a decade or two after the death of Jesus, and thereafter every action and writing of the followers of the way of Jesus, save perhaps a few of the Coptic Christians and gnostics, reflects Pauline doctrine, a huge overlay dwarfing and profoundly changing the original Jesus event.

Paul's Christianity adopted the accusation of sin and the selling of its antidote as the principal ways of spreading the word—a word based on guilt, shame, and punishment. Already in the Acts of the Apostles cultural intellect is regaining control, making its inroads into the guidance of spirit. Already the direction toward which the cross pointed is being reversed until the way that it offers, the crack in the egg, is eventually sealed shut. Paul's pontifical judgments, his convoluted intellectual analyses arising from his copious opinions on every issue presented in the new communities of believers, both instigated and completed the reinstatement of the cultural imprint that held sway before the coming of Jesus. And Paul has been quoted endlessly for two thousand years, with all the citations of him by Christian authors and preachers far outweighing the few comments given us by Jesus himself.

THE WANING OF THE PARACLETE

▼

Even more critical than other reversals of Paul was the dissolution of the Paraclete, which should have evolved into the most powerful field effect in our life. Paul's judgments replaced in one blow our opening to, direct contact with, and sole dependence on that wisdom of the heart Jesus brought about, the intelligence that is our ever-present friend, companion, helper, and inner guide. Paul himself set out to be the chief guide, the supreme authority in matters of the spirit, rendering the spirit within us superfluous. His letters are filled with endless justifications of his own authority and assertions that, in spite of not having known Jesus, he was the equal of those disciples who had. Paul introduced to history the notion of an isolated self demanding justification. How to justify ourselves in the eyes of this jealous God becomes paramount, and justification by any means becomes the backbone of Christendom. The good Father who judges no man and rains on just and unjust equally, the God with whom no justification is needed, is forgotten.

It is easy to see how, in those early years of Christianity, the foundations of authority shift from the Paraclete and individual—our own heart and mind—to Paul, and from Paul to the long parade of elders, deacons, bishops, and popes. Finally, adorned in its robes, occupying high places, seated at the heads of tables and then nations, this supreme—if not quite divine—authority is backed by mighty armies, makes the decisions and choices, and shapes the opinions of its followers, guiding them like lost sheep. With this, we are back to zero. The gospel is dead. Long live the Church, Creed and King, Caesar, Pope, Emperor, right down to the holy global economy. The Christian culture is born.

Ludicrous in their irony are the many ways that this culture was not merely reinstated but also strengthened through the cross. The one who said "call no man father save that Father in heaven" is drowned out by centuries of self-declared fathers who, once all the bloody battles among these groups had arrived at some sort of stalemate, rushed about to be called "father" by their flocks, granting degrees of fatherhood on each other like academics granting twentieth-century doctorates. And the one who said the time had come when worship of God was in spirit and truth rather than in temples or on mountaintops would have been amazed to see temples sprouting like mushrooms and worship services taking on the pomp and circumstance of coronations or grand opera.

The irony of so many of the sayings of Jesus undergoing reversal by the institution of Christianity is only the outward effect of a far greater betrayal within: the reinstatement of judgment and retribution alongside if not in the place of love, compassion, and forgiveness. And the issue of reconciling these two irreconcilable positions, love and law, has tied Christians into intellectual knots of apology for two millennia as they continually rationalize away the inconsistencies and fabrications that arise from trying to make old and new agree.

Some three hundred years after the time of Jesus, the whole crazy quilt of evangelical hucksterism was brought together as the original gospel, a New Testament of man's relationship with God. To insist, after this long and messy process of compilation, that this strange hodgepodge was the infallible word of God, complete and without contradictions, and, further, that it was to be accepted literally, surely required the most superb "knight's move" logic, the kind that could skip over syllogisms with ease. But those who promoted this gospel did not stop there; they declared that belief in and complete acceptance of this polyglot was absolutely imperative to the salvation of our soul.

MONUMENTS OVER THE MURDERED

▼

Gil Bailie and René Girard wrote brilliantly on culture surviving through murder. War is, after all, organized and religiously sanctioned murder, as is the death penalty. Indeed, imprisonment is a macabre, government-sanctioned form of slow execution spread out over years.

One definition of a *prophet* is a person who threatens culture's power structure by holding up a mirror to its folly and showing where such folly leads. Jesus observed that culture kills such a prophet, and, having killed the prophet to be rid of his threat, that culture then builds a "monument over the prophet's grave." These monuments are the constructed mythologies through which prophets, once they are safely dead, can be converted from cultural critics into cultural supports and made objects of saintly hero worship to serve culture. Jesus obviously intuited that this travesty would be the outcome of his own gesture, but of course this couldn't deter him from playing the hand destiny had dealt him.

Consider how American culture first demonized and killed off the native peoples who were here when we arrived—we stole their land, and then, once they were safely out of the way, built monuments over their graves. Through mythologizing we reversed our demonizing and read into those murdered peoples a history of great spiritual teachings, legendary heroics, and saintly nobility of character. Easing our national conscience concerning our murder of their forebears, we dumped on the stray survivors of our ethnic cleansing the heavy burden of a glorified imagery few of us could live up to. While we attributed to them pseudo-religions and spiritual paths, giving them a larger-than-life nobility, we also succumbed to our cultural impetus by selling the symbols and myths we created. We hawked the false wisdom of counterfeit medicine men and women on the workshop circuit and fed like vultures on the corpses beneath the monuments we had erected while our treatment of those who remained was as shabby and shameful as our eulogies were overblown and hollow.

American culture killed Martin Luther King Jr., and then, once he was safely out of the way, built a monument over his grave as well, making a saint of him, naming streets, boulevards, schools, and institutions after him, at the same time allowing the condition of his people to steadily deteriorate under new words covering the same old cultural travesties. Political correctness, while seeming to promote racial sensitivity, is an agreed-upon form of social lying in which the most biased and prejudiced among us can unctuously say the proper words and thereby cloak our continuing destructive patterns.

The God of love, a long time in coming and pointed toward by Amos, Isaiah, and the Psalmists, found its culmination, greatest spokesperson, and ultimate model in Jesus. His whole address had been to his own people, whom he obviously loved enough to give his life in the hopes of lifting them up as he himself had been. As with the long line of great prophets before him and as it would be for those who came after, however, the cultural power structure that killed him then built the monument called the Christian religion over his grave. Our pointing fingers to identify the good guys and bad guys involved in his death—Jews, Gentiles, Romans—is all froth and beside the point. The real accuser and murderer was culture; the motive was the preservation of its power and mind-set.

Ironically, the new religion that emerged after Jesus' death found the fuel to feed its flames through pointing the finger of guilt back toward those who gave rise to Jesus and for whom he died: his own people. This is a common trick of all revolutionaries and their revolutions. The new religion could survive only by creating a demonized enemy rooted in its own origins and onto which culture could project hatred and rage in order to organize and galvanize its followers—often into violent actions.

Not only was Jesus the target of just such a cultural power struggle, but also the people caught in that struggle ultimately became the target for the new religion woven around him. It is plain to see, then, how it is that the Jewish people rejected Christianity. The same cultural cycle of murder and glorification of the murdered simply turned again. In the hands of the literal-minded evangelists, however, the lofty heights of Old Testament thought became a travesty through which the Jewish people, along with mankind as a whole, lost the best of both worlds: the Jewish people the light of their greatest prophet, the Christians the light of the Old Testament. That wondrous collection of magnificent love songs, fables, historical myths and legends, psychologies and philosophies, profound prophetic foresight and spiritual insight gathered over millennia simply can't be read on the literal level attempted by New Testament chroniclers—and still attempted by fundamentalists—without irretrievable loss and much misunderstanding.

The strength of the gospel nevertheless worked among the Jews as elsewhere. Because their theology was the very root of the gospel, the gospel was the high point of their own history. We have only to examine the long line of great mystics and saints arising from the Jewish people—the great Bal Shem, and Martin Buber in our own day, to mention but two. And let me mention here that any nation expelling its Jews, as did Spain in the Renaissance, generally enters an intellectual, artistic, and probably spiritual dark age. As of today, one third of all American Nobel laureates are Jews, who represent 3 percent of our population. The Holocaust may have been a darker sign than we have yet comprehended.

THE CHURCH AS MEDIATOR

▼

Christianity turned Jesus from our evolutionary model into the greatest tool of culture. Converted into the christ, Jesus became the Great Mediator. No longer the model of higher development, the one who draws us toward him through lifting us up, Jesus as the christ became a go-between, mediating between the wrath of that same old tyrant Jehovah and the same old sinful, victimized, and helpless human.

This additional mythical creation, the Great Mediator, was itself mediated by the church that invented the idea. The church became the mediator between an individual and his or her own spirit—a double mediation or double indemnity. Or a double cross. With this new and powerful role, an extraordinarily efficient means of cultural and social control was instituted.

Here, in this new role, was an institution imposing its mediation between an individual's heart and brain—an invasion of our very biology, a violation of the single most intimate aspect of evolution's great venture into consciousness, and precisely what Jesus objected to in the actions of the Pharisees and Sadducees. Above all other things that Jesus observed was that there could be no mediation between an individual and the kingdom within, the heart of God, and he showed real anger at those standing at the door and not letting anyone through—ironically, precisely the cultural effect now smoldering in fundamentalist Moslem, Jewish, and Christian movements. This final assumption of mediation, however, opened the door to centuries of knaves duping fools as Christianity stumbled from the irrational to the irresponsible.

One of the miraculous strengths of the gospel, however, lies in the simple fact that in spite of all this, great and noble geniuses of the spirit arose continually out of this strange paradox, and still arise today. That steady stream of great and noble women of the church, in spite of the political and economic antics of the papacy and its Protestant counterparts, quietly did the will of Jesus' father, tending the poor and dying, doing what could be done to repair the damage wrought by the great cultural powers. The names of these women are legion, and they still quietly go about their work today while theologians wrangle.

"BY THEIR FRUITS YOU SHALL KNOW THEM"

▼

Not long ago—in fact, during the writing of this chapter—the pope of the Catholic Church apologized for the Church's past misdemeanors. In fact, the pope had declared the entire last year of the old millennium a time of atonement, for reflection upon and confession of wrongdoing. Though the pope did not elaborate on the Church's own acknowledgment of wrongdoing to any extent, his quasi-contrition was without precedent in Church history.

The most powerful criterion for behavior ever conceived is found in the simple statement, "By their fruits you shall know them." There is no judgment implied in this admonishment by Jesus, but it is the one criterion that none of us, and surely no institution, can tolerate. Know us by our public relationships, slogans, statements of policy, mission statements, lofty ideals, creeds and beliefs, confessions of faith, brochures, proposals, and public apologies—but not by our fruits, results, actions. Even if an action were to be suspect, the institution as a whole is never held at fault—it's just some bad characters in the ranks, human nature, you know, rotten apple in every barrel!

The word *religion* comes from the Latin *religare, re* as in re-turn or re-peat, *ligare* meaning "to tie, to bind," as in *ligation. Ligature,* "the action of binding," comes from the same root. Perhaps what we need is not to re-tie with a mythical past but to move on with our new findings concerning heart and brain. Religion is often associated with tradition, and *tradition,* like the word *traduce* (to betray), comes from the same root word meaning "to transfer," or "to give over," as to a set of laws. *Trade* comes from the same word, a giving over of one possession for another. To give over to something or someone is to bind oneself, to surrender some part of one's being for something outside one's self—an inner betrayal.

Gil Bailie related this to the underlying meaning of the word *desire:* "to want something outside ourselves." He points out that desire is destructive when mimetic or imitative, as when we desire something because others have it and we don't. In such a case, we want something that is not actually of our own being, but costs something of our self to get. This is the faculty on which television, the Internet, a global economy, and industry depend. Desire is not longing. We long for that which we sense is within us but seems unavailable to us. Longing is a gift.

Jerome Bruner spoke of the great power of language to pass on to us the knowledge of the ages—of tradition. We believe this implicitly and to question it seems silly. Recall that Suzanne Langer claimed our greatest fear was a "collapse into chaos should our ideation fail us." Culture, however, is tradition, an ideation shaping us at the cost of our spirit and freedom. Tradition, then, can be bondage and devolutionary, but it becomes a field effect like any other, shaping each mind born within that field's influence. We sanctify and defend our inherited field with passion, for it shapes our very minds.

Jesus urged us to renounce all crippling ties, saying, "Let the dead bury the dead." Krishna, his nearest counterpart in the East, took the same stance in his dialogues with the hesitant Arjuna facing his destructive relatives, who, Krishna reminded him, were already dead.

Use of language to pass on a cultural heritage can pass on chains forged over millennia. In subtle ways language, culture, and heritage can give rise to and reinforce each other. Jesus' way broke with culture and bondage, had no tradition, and hasn't one today. His way forms anew for each one of us taking it on; each person picking up the cross is part of a particular union that has not existed before.

We have scant modeling or instructions for transcendence in our day. Culture feeds the ancient survival modes of our brain and keeps us locked into them. The gospel countered these cultural chains until the church created its own gospel based on its accusation of sin—and we couldn't hear the gospel of love for the noise of the Doomsday trumpets.

Even in its guise as advertisement, textbook, and school, state-religion continues its accusations, suggesting that we are guilty of incompleteness, lack, separation, not belonging, and/or of being cut off from God. From conception, enculturation is an automatic, cellular implanting of this conviction of sin, and one that is self-fulfilling.

Whether or not we have any kind of religious affiliation or religious belief makes no difference. The Christian accusation of sin is part of the very fabric of our culture, and the more subtle its presence, the more powerful its effect. It underlies our whole legal and legislative fabric and convinces us that without law and its justice, society would run amok just as it convinces us that without harsh prohibitions, our children would do the same.

To suggest that we are not guilty of anything, that our children are perfect as they are, that we would not turn to murder and mayhem were the long arm of the law not omnipresent, or that all our needs would be met by a benevolent nature, as seems to have been Jesus' position, can be a major cultural heresy today, as in Jesus' time. As chapters 6 and 7 articulated, to reject the accusation of sin undermines the foundation of culture and its church and schooling. So to call culture's great belief—that without enculturation humankind would be beastly, primitive, and dangerous—nothing more than a lie is a major heresy of our or any age.

Saint Thomas Aquinas, the great Church Father of the high Middle Ages, wrote a fiery discourse addressing heresy in which he specified why heretics should be burned at the stake and thereby gave his sanction to that strange carnage of witch-hunting. Indeed, English historians estimate some nine to eleven million women and a sprinkling of men were so dispatched in the centuries following Saint Thomas's dictum. History is largely a fiction, according to Will Durant, but there is evidence that a great deal of murder took place over those centuries, and that women were the chief target.

By virtue of their very nature, women are somewhat automatically heretics or challengers to a reigning male power structure in which intellect and the interpreter mode are absolute. The presence of a strong woman perceived as challenging can enrage any male, particularly an avowed celibate male who is supposedly immune to the sins of the flesh. The combination of violence, power, and lust is not a recent phenomenon.

Some of the most tragic yet stunning dramas of history have arisen from this combination and the struggles that ensued. The martyrdom of the Beguine Marguerite Porete, burned at the stake in Paris in 1310, stands out in my mind. We have her little gem, *The Mirror of Simple Souls,* in spite of the attempts of the Inquisition to erase all traces of her existence. The accounts of her quiet, powerful demeanor during her long imprisonment and horrifying execution reflect the cross in every way.

Again, the simple observation "By their fruits you shall know them" is the one no power system can tolerate, above all the institution of Christianity. One burned heretic or drowned crone shoots down the house of cards, whether the house be Catholic or Protestant. And where in history can we find the equal of that arch misogynist John Calvin? Parading from village to village with his small well-armed army of Protestant inquisitors, Calvin's search for witches outdid that of the Catholics, and for crude, barbarous vulgarity is unmatched anywhere. Calvin overwhelmed the peasant farmers with their pitchforks as they tried to protect their women, and forced the women to disrobe so that he might find any trace of the infamous "witch's tit." (It seems that beneath the Puritan fanatic there existed the puerile voyeur.) Drowning and burning women, sometimes en masse, he went about preaching his gospel of predestination, sin, and death. (And we revere him today as the founder of the Presbyterian Church—the businessman's church back in my childhood, perhaps because it offered sanction for wealth predestined by God.)

Martin Luther, having declared every man his own priest, turned on the peasants with his own army when the peasants revolted against their oppressive landholders and crushed the ensuing chaos with a bloodbath of no small order. And that Bible-bearing, gun-toting purifier of culture Oliver Cromwell tried to eradicate not only all forms of art (which, said Blake, is the first act of Satan, followed by the removal of pleasure, and leaving only grim necessity), but also all signs of heresy, such as, ironically, Catholicism, whose followers were prime heretics in his book. The march of Cromwell's stoic, hymn-singing Roundheads left as horrible a wake of murder and pillage in Ireland as can be found in history, hardly eliciting a love of the English in that torn land.

BITTER FRUIT

▼

Consider in our present day the practicing Christian lawyer, judge, jurist, policeman, jailer, Army chaplain, officer, bomber pilot, soldier, politician, business tycoon, abusive parent, women's clinic bomber, patriot with Bible in one hand and gun in the other. Picture them continuing their honorable professions while carrying the cross, turning the other cheek, following Jesus' way. Consider our two million brothers in U.S. prisons, their number doubling each decade, and the growing numbers on death row. Then picture each of us acting with compassion and forgiveness instead.

I recall the newsreels when I was a child, showing the pope blessing the Italian army on its way to bomb, gas, and machine-gun the spear-wielding Ethiopians (whose country, ironically, as home to the earliest Coptic congregations by the end of the first century, was the first Christian "nation"). The pope had a long precedence for such benevolence: The eleventh-century Pope Urban the Second cried: "God wants it! God wants it!" as he blessed the knights on their way to kill the Moslems in the Holy Land and, at the same time, set a precedent for future mass murders. And we too had our own Cardinal Spellman, who blessed the troops on their way to "Christ's Holy War" in Vietnam.

As I mentioned in my preface to part 2, when I turned eighteen I enlisted to fly and fight with the Army Air Corps in World War II. The few pictures of us at that time show how astonishingly young, indeed childlike, my friends and I appeared—and were—friends who went down right and left in a carnage that cost upwards of thirty million lives, a worldwide nightmare fomented, instigated, and carried out by the two greatest historic strongholds of Christendom itself. The homelands of gentle Francis, Eckhart, Tillich, and Bonhoeffer were also the homes of the Third Reich and Fascism. The Holocaust they instituted will surely stand as the most hideous of all human nightmares, the cross reflected seven-million-fold. More than twenty million deaths in the straight slaughter of war was one thing, but the cold, deliberate mass murder of those millions of Jews, following their prolonged nightmare of torture, humiliation, degradation, and pain, carried out by a Christian nation stands as a final witness to travesty. By their fruits you shall know them—this is hardly just a few bad apples in the barrel. And the silence of the papacy during this horror is not quite assuaged by a generalized apology coming a safe sixty years later, public relations cameras cranking away.

The carnage has hardly ended, of course. There has been the mutual strangulation of the Christians and Moslems in Bosnia and Kosovo. In

Catholic South America, particularly in Brazil and Colombia, there are an estimated nine million homeless, abandoned children between the ages of four and eleven living beneath the streets at any one time. When they come out at night to search for scraps, they are systematically hunted down by the police, stacked up like cordwood in trucks, and carted off for mass burial before dawn, as Thom Hartmann has described (see Thom Hartmann, *The Prophet's Way*, Mythical Books, 1997), while the pope pontifically condemns birth control. "This staunch and virtuous position is necessary," I was told by a Catholic father, "to preserve the sanctity of sex."

Just as marriage—a cultural institution—is generally a disaster for all but the strongest and most enduring relationships, religion—another cultural institution—has been a disaster for humankind's relationship with God. Surely both deserved something better. The cross was the attempt to bring new life to that relationship and Christianity was culture's means to nullify that attempt, culture's "homeopathic remedy for the viral threat of Jesus."

Yet great saints emerged and still emerge in a steady if small stream, from the system and in spite of the system. The question is, would they not have emerged from history anyway, whether or not an institutionalized Christianity existed? And the answer is: Of course they would have. "Even these stones would cry out" the good news that Jesus brought into the world. What if only the love of God and our indissoluble union with him, as manifested in Jesus, had been broadcast to all nations without the interference of a false mythology and the concepts of sin and guilt? Love can only offer itself, and can lift us up only if accepted. Recall Gil Bailie's eloquent assertion that the cross depicts the utter powerlessness of a God of love. What happened to that God? Where did he go? The accusation that the church has kept him hanging on that cross these two thousand years is not too far-fetched.

The early Renaissance painters discovered perspective and in doing so changed the perspective of humans in general. The image of an innocent love of God—an image sacrificed to law and culture—would have brought a slow but steady change in human perception everywhere, as it did anyway, in spite of the accompanying shadow cast by the institutions and evangelists. Creator and created had found a new name and face in Jesus and his father, and that evolutionary leap was bound to win out.

We don't need church under any brand name, with its accusation of sin and selling of redemption, its huge bank accounts and real estate, lawyers and lobbyists, political games and public relations, radio and television stations. We do need that steady stream of selfless people, particularly women, the church has given us, in spite of itself. People like Peace Pilgrim simply materialize to exemplify the gospel. They crop up continually, if always on the fringe, always suspect by the authorities and respectable church people. Little Second Comings occur all over the globe.

There are no mediators between our heart and mind, just the blocks of our defenses, fears, and doubts, as Peace Pilgrim clearly displayed. So we might as well take the chance, quietly and without fanfare, not for public display but in our private place of heart. We might as well take the leap and drop defenses, judgments, the fearful passion for prediction and control, the dreadful need for self-justification, with no thought of tomorrow. This is that simple, private move the gospel offers us—picking up our cross. Rumor has it the burden is light.

PART THREE

▼

BEYOND
ENCULTURATION

▼

ADVENTURES OF SPIRIT AND TRUTH

In my thirty-second year, now married and with three children, I sat in my office in silent, wordless contemplation one day and felt the presence of my hero and model strongly. I fell out of my body into a vast ocean of quiet that left nothing to report once I returned to my ordinary state. This was not an experience of ecstasy or great enlightenment but was instead an immersion into a deep, dark sea of calm. Thereafter for some three years I moved in a fluid drive wherein everything worked to perfection, with almost no effort on my part. I knew myself and family to be cared for, nurtured, and intimately loved and protected by some deep interior presence.

This fall into grace was preceded by an exciting period of discovery. I read Paul Tillich's mammoth *Systematics of Theology* in its entirety, a huge trilogy my department head confessed he had despaired of reading but that I found both a challenge and a delight. I read every available work of Soren Kierkegaard and wondered how experiencing his book *Purity of Heart Is to Will One Thing* could be at once the most soul-shattering, unhinging event and an exquisite intellectual, aesthetic, and spiritual feast.

Besides reading, I began working on what was to become, twelve years later, *The Crack in the Cosmic Egg*. The book unfolded with the same sure coherence and meaning that characterized my life in general during this period. It was about this time that I discovered the journals of George Fox, founder of the Quaker movement, and his practice of opening to the spirit, which I realized was roughly parallel to unconflicted behavior. Fox's practice of opening works for those who have the patience to wait without doubt. When I was stuck on a particular problem in my writing or felt like

I was at a stalemate, I would go into the library of the college where I taught and simply stand and wait. When a certain detachment took the place of my ordinary, chaotic internal chatter, a pull would come, which was as powerful and sure as though I were moved by invisible cables, and draw me to a particular section of the library. I would find myself reaching to some obscure shelf for some equally obscure book that would literally fall into my hands. I would automatically turn to the page that contained precisely what I needed to know at that time in order to be able to move on in my writing adventure. Probably 90 percent of the references used in that book came to me in this way, most notably the discovery of Carlos Castaneda at the time of the release of his first university press printing in 1968. I was making the final edits of my manuscript when I was led to his book. It lay with other new arrivals on a table in a back room of the library, waiting to be cataloged by the librarian. A chill ran up my spine as I picked up the little volume, and the presence of the uncanny, the inexplicable, filled me. That I was led to this first book of his was a demonstration of what both of our books were about—both Castaneda's perfect gem and my rough manuscript examined what happens when we drop intellect and let the intelligence of "the other" take over. I realized later that the other was simply another term for the intelligence of the heart or the Paraclete spoken of in the gospels of Jesus.

In 1965 disaster fell with the death of my wife, who was only thirty-five years old. Her illness involved spontaneous healings and relapses and a series of paranormal events that would require a book of their own to detail and clarify. With the help of the minor miracles she brought about after leaving us, including several visible visitations, I held our four young children together as a family, maintained a teaching position, and kept my book alive and growing, more or less.

It was some two years later that the most significant and intense mystical experience of my life took place. It occurred one evening when I was quite awake and began with the slow but complete materialization of my long-lost anima figure. She manifested in my arms, in full, tangible, physical form, her lips and body pressed against mine. She was now, however, a composite—my adolescent love surrounded by the presence of some ancient, archetypal *She* who was simultaneously the feminine Shakti and the power of creation herself.

While our previous combining had been one of spirit, more or less, this time the fusion with my anima included our actual bodies, cell by cell, with each fusing cell a complete ecstatic explosion unto itself leading, finally, to our collective fusion with that vastness that has no name and defies description. Love is too hopelessly abused and inadequate a word for the state I experienced, but I have found no other. I am left silent, for as Eckhart said, all names and words must be left behind when we enter that cloud of unknowing, nor are words applicable in any way to that state once we are outside it again.

The intensity of this event eclipsed any and all of my spiritual experiences before and after. But the eventual slow separation from both the vastness and my anima—rather the shedding of the greater part of my self, piece by piece, until I was finally left in my ordinary state again—was a devastation that nearly unhinged me. Only my responsibilities to my children and my model, Jesus, held me to the near-barren life that remained.

In the years following this I had many experiences resonant with that major one, and these grew in depth and frequency. But some part of me seemed to have been cut off and lost and I have never again known anything equal to the enormity of that greatest fusion when I was forty. Later I was to read the remarks of Bernadette Roberts about "breathing the divine air" for weeks and the shock on having to come back into the ordinary world; she spoke unabashedly of finding this earth a living hell by contrast.

In 1974, working on my third book, *Magical Child,* I hit a snag in writing about the issue of child play. Why do children want to play all the time, practically from birth and to the extent that they will turn everything into play if allowed, while we adults have a totally different agenda for them, one we too have held since our own departure from childhood. We are convinced our agenda for our children is good for them and are perplexed when they resist this apparent benevolence with all their might, mystified as to why it is that nature would build into children a compulsion to play when responding to our adult wisdom about learning to survive often seems to suffer as a result. I gathered all research and studies of play I could find and sorted and synthesized, trying to find the answer among the hodgepodge of theories.

I spent several months on the question, getting nowhere, until one evening I felt the answer to be right there at my fingertips, ready to break through at any moment. I grew excited and spread all my various research notes into categories before me, knowing the answer was there if I just had the strength of mind to hammer it through. Finally, long after midnight I leaned back, exhausted, my enthusiasm spent and nothing accomplished as the answer eluded me yet again. In the most genuine and spontaneous prayer of my life I called out, "Oh God, what is the purpose of play in our life?"

With no warning, a wave of energy swept my body from my feet up and I found myself physically propelled at tremendous speed into an infinite black space slowly populated by a universe of unending stars and galaxies. I was tossed again and again, gently and playfully in an exhilarating but helpless fashion, much like a juggler tosses a ball or a father his child, from one end of this vastness to the other in joyful exhilaration. This ecstatic experience went on and on and I found myself shouting out, over and over, "God is playing with me!" After what seemed an endless time, the event slowly wound down, the starry space dissipated, and my room formed around me. I wept for a long while afterward from sheer gratitude, awe, and wonder over such a gift. From that point on, however, I knew that play was the whole reason for and essence of life—and not just for children, but for all God's children, whatever their age; and I understood that our great model's observation that we must become again as little children meant precisely what it said. But I also knew that our refusal to play and our prevention of play in children, our insistence on forcing them into defensive procedures, were evils of long standing. I had no idea how I could put this into words in my book, for the issue was overwhelming and seemed to dwarf my paltry attempts at expression. But play shaped the final drafts of *Magical Child* and entered into the makeup of all my subsequent books. Later I was to meet Michael Mendizza, whose Touch the Future Foundation gathered together many of those who had realized that play was the answer: Fred Donaldson, Stewart Brown, Chuck Hogan, each interpreting play in a different light—as was right and fitting.

Yet play was still fundamentally an intellectual concept to me and I realized that intellect was largely the shaping force in the play most institutions and groups fostered, in spite of all their good intentions and efforts. Not until some ten years ago, when I learned about and entered unconditionally into the physical expressions of Education Through Music, or ETM, the brainchild of the late Mary Helen Richards, did I rediscover play in its original state. ETM is an experience and can be known only through doing—thus, because it defies description, the Richards Institute faces problems with becoming well known. Nearly anything said about ETM misrepresents it. I can only claim that our body's knowledge of what play is and how it opens us to unconflicted learning can be rediscovered through the movement and singing activities of ETM. Through the physicality of ETM the last shreds of defensiveness dropped away from me and I discovered true play in myself—I could understood fully what our great model meant when he suggested becoming like a child. I also understood that no intellectual concept or mediating, mental effort can open us to this state. If our children's schooling were to include ETM, play as a force could be opened again in our children—as well as in adults.

Some three years after finishing *Magical Child,* a reader sent me a book called *Play of Consciousness,* by an Indian meditation master, Swami Muktananda. I was struck by the title but had no use whatsoever for the flood of gurus swarming our shores in the late 1970s and I felt something akin to contempt for the smirking and rather sinister-looking character on the jacket of the book. I had my model for transcendence, had lived with him all my life, and didn't need another. That night, however, I decided to at least glance at the book, having nothing better to read at the moment.

I had lived with my new wife for three years in a remote section of woods in the Blue Ridge Mountains of Virginia, two miles from the nearest road, electric line, or telephone. A two-thousand-acre farm lay abandoned to the east of the cottage I had built for us and a two-thousand-acre uncut forest owned by a lumber company lay to the west of us, and the forest tract had no access except through the vast, empty farm itself.

By the light of my trusty Aladdin kerosene lamp—the equivalent of the light given by a forty-watt bulb—I sat down that evening to see what kind of wool this particular Muktananda charlatan was pulling over the gullible eyes of his readers. I opened the first page, read the opening sentences or so—and the curtain fell. For the fourth time in my life that astonishingly huge weight lowered, pushing me out of all body consciousness, as it had some thirty years before. After the briefest moment, however, I was pulled back into body awareness by a brilliant light shining in my eyes, and I opened them in complete confusion.

As I've said, our place was remote—and difficult to reach. Two locked gates barred access to each of the many fenced-in meadows that had to be crossed to get to the woods where our cottage stood, fairly well hidden even in daylight. How could a car have gotten in, and why would it be throwing its spotlight through the window, right into my eyes? Such questions occurred to me as I opened my eyes and strained to see through the brilliance.

There, inches from my face, was a white alabaster bust of Jesus, complete with pedestal holding it. The light was streaming from the white marble and I glanced at the statue's eyes. They were brilliantly alive, looking at me intently, when without warning the statue leaned over and blew its breath up my nostrils! As that breath filled me I fell yet again out of this body and world into that vastness that no words can describe, where words, names, and being must be left outside.[1] The episode was short but decisive and to the point. Once back in my body and world, I put aside that explosive book and commented to my wife that wherever this Baba Muktananda character was, we were going. I was later to learn that Baba occasionally gave *shaktipat*, the passage of spirit or power from teacher to student, by blowing up the student's nostrils.

Some six weeks later we sat at Muktananda's feet in his ashram while his designated successor, Gurumayi, interpreted for him, because he spoke no English.

I blurted out to him, as an opener, the peculiar circumstances that brought me to him and added: "Baba, I think you are Jesus."

Muktananda laughed with great glee at this, slapped his knees, then, pointing his finger, looked me in the eyes and (through Gurumayi's interpretation) said, "Why, of course! And so are you."

We stayed on at the ashram and within the month I twice experienced that great weight suddenly upon me and pressing me into another awareness. Each time it wasn't to be wafted out to the nether regions in ecstasy, but taken inwardly, mentally, to be literally taught some aspect of Baba's cosmology based on Kashmir Shaivism, which I was unable to grasp intellectually. In that knocked-out state I would live into, tangibly experience, some particular didactic teaching, and then know in the very cells of my body, not just in my head, what Muktananda wanted me to find out. Being of a slow-minded nature, I understood little of what he told me directly and less from trying to read those intolerably boring Sanskrit scriptures he insisted I tangle with and which put me to sleep nearly instantly. So this

---

1. The significance of this symbol is hardly subtle: Here was my long-loved hero, Jesus, manifested as a marble statue with eyes that were brilliant and alive, and who blew up my nostrils that which could only be the Holy Spirit itself. And while Jesus was my idol on a pedestal, here he was embodied—incarnated, I was clearly being told—by a very real human presence among us.

temporary knockout was simply a shortcut procedure he used to get me to understand. (In Siddha yoga they called it *passout* meditation and I have experienced it several more times over the years.) The similarity with my anima's technique to shift my attention more than thirty years earlier was not lost on me. Cosmological principles are the same regardless of context. Creator-created dynamics can reflect on and use whatever content our reality offers.

Siddha yoga meditation centers on the heart, and I learned almost everything I would ever know about the heart through this discipline. At one point I had about three hundred sayings from Baba and Gurumayi concerning various issues of the heart, all of which have been borne out since in actual research. Baba had explained to me in the beginning of our relationship that within ten years or so the scientific community would present all the research needed to explain what he told us and what we directly experienced regarding the heart in Siddha yoga. Over the years I continually received research papers on the heart and eventually I discovered, through the Institute of HeartMath, the rich new medical field of neurocardiology, all of which proved Baba's prediction quite valid.

Muktananda spoke many times of a subtle sphere of energy surrounding our body like a cocoon and referred to it as the vibrations or wave forms of Shiva, the primordial god of the Hindus and a major force in Kashmir Shaivism. He claimed that these wave forms were the frequencies out of which the universe formed and that the subtle sphere surrounding us contained the whole universe within it in subtle or potential form. I had read Carlos Castaneda's description of the luminous "egg" in which our life takes place and figured the two were the same.

One memorable evening in the the ashram I was seized with the notion of doing *pranayama*, the "breath of fire," until, I vowed to myself, I had a breakthrough of understanding about these vibrations of Shiva. Pranayama is a rapid deep-breathing meditation that I had been warned to do very lightly because it was supposedly hard on weak hearts and older people.

Being of even weaker mind than heart, I threw caution to the wind and decided it was time for a showdown with my spirit. Dripping with sweat but with continually renewed energy, I did the breath of fire on and on into the early-morning hours. When this breathing, which had become automatic, suddenly stopped of its own accord, leaving me in great stillness, I clearly perceived the subtle sphere of vibrations engulfing me. It was so palpable I felt I could both touch and embrace it. The word *plasma* came to mind for this sphere that was like a living presence, and I realized that this truly was a cocoon in which we were immersed all our lives. To me it was pure love—vibrant, alive, totally nurturing. Within it was the entire universe in its subtle form, or "implicate order," as physicist David Bohm might have called it. Later, when I read about the electromagnetic fields emanating from our heart, I thought: "That which we call a rose / By any other name would smell as sweet." My cellular knowing had been given me first, a frame within which this intellectual information made very good sense.

A year or so later I experienced this same subtle sphere as a force field revolving around my wife as she gave birth to our daughter, there in our bed in our little house in the woods. She had been kneeling for quite a while doing what we knew to be pranayama, but which was called the Lamaze childbirth method, when I, trying to be of some use, thought of supporting her back a bit. But she was surrounded by a field of such energy and power that I bounced back from it like a Ping-Pong ball, as I later described it to her. She explained, in turn, that she simply needed to be alone at that time, to do the work at hand. Our daughter was born shortly after my attempt to help.

I was in Siddha yoga for twelve rich, rewarding, fruitful years. My wife, daughter, and I spent nine winters in the ashram in India and summers in the ashram in America and various ashrams around the world. During this time I gave nearly two thousand talks around the globe on the mind-heart connection and nature's plan for our development. After a decade or so, my inner promptings told me it was time to leave that way of life, but I loved it so much I ignored these messages. Finally, in 1991, in preparation for a talk I was to give for the guru's weekend intensive, I was reading Bernadette Roberts's *The Experience of No Self* and came across her cryptic remark that no one can get anywhere in the spiritual life without a frame of reference, but that always, at some point, we must leave our frame of reference.

This notion disturbed me but I dismissed it and, as planned, gave the talk I had been preparing. Toward the end of my presentation, though, the color seemed to drain out of the great hall and the audience before me, and I knew I had left Siddha yoga. More accurately, I knew I had been thrown out by some force within me. Without missing a beat I completed the talk, realizing it would be my last, and left quietly, without a backward glance, without fanfare, deeply grateful for the privilege of all I had experienced and learned—

which was vast, considering where I had begun, and which I still feel today.

A couple of weeks later, as I sat quietly at home, musing over what might be next in this marvelous adventure, I felt a stirring within me and heard for the first and only time an audible inner voice. As clear as if it came from someone sitting right next to me I heard from my heart the words: *Will you give up all support systems and follow only me?* Instantly and emphatically some deep part of me all but shouted, "Of course!" as though it would be ridiculous to consider anything else. I felt rather left out of the discussion going on within me, but with those affirmative words *Of course!* I experienced an explosive and ongoing shower of what in Siddha yoga is called the Blue Pearl, a tiny point of intense, electric blue light that just appears at random in our visual field.

Until I met Muktananda I had never heard of, much less experienced, such a visual effect—it is referenced in American Indian ceremonials, as I found out, as an affirmation of or granting of spiritual blessing. After I received shaktipat, the passage of power, from Baba, the Blue Pearl had popped up almost continually in my life, particularly after some rewarding effort or as an affirmation of right action. And from the moment of that affirmative *Of course!* in response to the voice in my heart, the Blue Pearl had simply exploded everywhere around me, a cascade of electric blue sparks showering down like rain, which continued for a half hour or so before slowly diminishing. I certainly don't feel that I have followed that inner voice particularly well—I have wobbled all over the place as usual, rather than nobly walking the straight path we idealize in our mind, which means that life has continued on as usual, as rich and as full as I can manage. My inner voice brought no blinding light of realization or eternal salvation or opening of my doors of perception. It did, however, bring a strong affirmation of life in general, for I knew even as it spoke that the voice from my heart was that of my long cherished hero, Jesus. With the Blue Pearl's fireworks, I also knew that it was Muktananda. And then I knew it was Meister Eckhart and Marguerite Porete and the Sufi Ibn Arabi, Theresa and John of the Cross, Francis and Jean-Pierre de Caussade, and all the other known and unknown in the throng that makes up that eternal Second Coming of, as a gnostic gospel offers, "He who is always becoming as we have need of Him to be."

On reflection I knew too that in some obscure way, as Muktananda had said, the voice was also my own.

Nine

▼

LASKI'S REVELATION

Man's perceptions are not bounded by organs of perception; he
perceives more than sense (tho' ever so acute) can discover.

—WILLIAM BLAKE

What accounting can we give for the appearance in history of a great being like Jesus that doesn't require some kind of frontal lobotomy to accept? In this chapter I will set up the framework for such an accounting, which may suggest at least a personal approach to putting the way of forgiveness and peace into effect.

In 1962 Indiana University Press published Margharita Laski's now classic work, *Ecstasy: A Study of Some Secular and Religious Experiences.* For me, her book was an ecstatic revelation in itself, and offers an answer to the above query concerning our great model.

Laski outlined a six-step process underlying *Eureka!* breakthroughs, those creative insights, revelations, and transformations of mind that change the history of science, philosophy, art, or religion. Her study validates the notion of fields of intelligence and information and offers valuable insight to the enigma of mind and creation even as her observations deepen these mysteries. The six steps she outlines in this creative discovery process are:

1. *Asking the question.* A suggestion, idea, or intuitive hunch, something we long to find out about, an experience we hear about and want to experience ourselves, an enigma we want to solve—whatever it might be, a quest must become not just the focus of our life, but such a passionate intensity that we are seized by it and feel we live only to serve it. Often we get an idea that we think will win us a place in the sun and serve our interests in the world. Until we are seized in the pursuit of our notion by

that which we think we have seized, the level of passion will not be sufficient to ignite our movement toward *Eureka!*

2. *Searching for the answer.* We must explore all avenues that might be useful in our search; find all the pieces to the puzzle; pursue every discipline; read every text; follow every directive whether the search is scientific, spiritual, philosophical, or artistic. Laski points out that we must leave no stone unturned in gathering the materials for our answer. The early stages of this pursuit are generally exciting, colored as they are by the conviction that the answer is always just around the corner.

3. *Hitting the plateau period.* A time of stagnation inevitably arrives when no more materials can be found, no more related paths can be explored, no more discipline can be harnessed, and no more sacrifices can be made. We have done as much as we can to no avail, which brings us frustration, despair, even bitterness and disillusionment—dark night indeed. Yet this is a time of gestation, that part of creation that lies beyond our doing and can't be engineered. We might go through several such plateaus before the answer forms and breaks through to us. And, of course, the breakthrough may never come at all.

4. *Giving up all hope.* No dawn follows the dark; all possibilities are exhausted; we have tried everything but no answer is found. We feel we have wasted our life to no avail—and we quit. Really. Period!

5. *Breaking through.* Real quitting clears the circuitry of mind, brain, and body and makes room for the answer to appear. *It* has access to us—at which point the answer arrives, full-blown and complete, when least expected, out of the blue, and in a single instant's insight.

6. *Translating the answer into the common domain.* This is the critical step, the one in which far more revelations perish than survive to see the light of day. Our instant, breakthrough insight can't be communicated as given—it must be translated into language that allows it to be shared with others of a like mind. Until this is done, it hangs in limbo, halfway between creation and created until it is properly birthed into the world.

Laski's six steps outline a clear way by which creator and created give rise to each other. To illustrate this I have chosen a few reports from a wealth of scientific, academic, and spiritual experiences, some of which I have used in my previous books.

LASKI'S SIX STEPS IN ACTION

▼

The Irish mathematician William Hamilton was seized with the notion of a quaternion function in mathematics and devoted himself to the issue with total absorption. For fifteen continually disappointing years he worked. Time and again he despaired and tried to put the issue out of mind, but then would think of a new approach and was off again. Finally, he truly quit, lamenting to his wife that he had wasted the best years of his life on a fruitless pursuit. Following this momentous decision to abandon his quest, he asked his wife to accompany him on a walk to ease his sorrow. As he and his wife crossed a little footbridge into Dublin, his mind bankrupt and blank, the answer fell into his head in a single blinding flash. He later reported that he knew at that moment that another fifteen years lay ahead for the translation of that symbolic flash of lightning into the cornerstone of modern math, his famous quaternions.

August Kekule, Belgian chemist, seized in a fashion similar to Hamilton, pursued a certain configuration in molecular structure following his intuition. The structure eluded him, however, no matter how he labored. Finally he withdrew, defeated, and sat by the fireplace, resting and drifting into a mindless reverie, when he perceived directly in front of him a snake with its tail in its mouth, an ancient symbol forming a peculiar configuration. *Eureka!* He had his answer. After translating his vision, he gave the world of science the theory of the benzene ring, cornerstone of modern chemistry.

Gordon Gould's experience was not only different from those of Hamilton and Kekule, but introduces an intriguing variable as well. Gould, an optical physicist, was at home for the weekend doing nothing in particular. Without warning there broke into his vision a symbolic structure of enormous complexity and detail, etching itself into his brain indelibly in a single flash of insight. He reported being "stunned, electrified" at the enormity of it and spent the rest of the weekend feverishly writing—page after page—the essence and remarkable implications of what he had seen. By Monday he had roughed out the theory of laser light, for which he would eventually receive the Nobel Prize.

The intriguing aspect of Gould's breakthrough is that he had gone through no preliminaries, no passionate search. Laser light was unheard of, and, unlike the search for the double helix discovered by Watson and Crick, for instance, was not in any way the focus of intense speculation and research. Gould was mystified over such a monumental possibility falling into his head unbidden. On reflection, however, he noted that he had spent many years studying physics and optics and for twenty years had been vigorously practicing that profession. Thus, he mused, he had been unknowingly "feeding into the hopper of mind all the materials, the bricks and mortar" for the awesome edifice that simply materialized and presented itself to him.

Suzanne Segal, an admirable person based on the one short account she left us, is an odd parallel to Gould. Her parents were Holocaust survivors and had little tolerance for matters of the spirit, an interest Segal displayed from age five. Receiving parental support for her spiritual longing, she discovered Transcendental Meditation when she was sixteen years old and took to it with passionate intensity. For six years she lived, ate, and breathed TM. She became a trainer, completing all the advanced work in Switzerland, and finally became a member of the inner circle of the organization itself. Before long, feeling saturated, stale, and disillusioned, she withdrew from TM and even stopped meditating.

For six more years she simply lived her life until, at age twenty-eight, she married a Parisian. Eventually, she became pregnant and, as the pregnancy advanced, began to experience strange, confusing perceptual shifts for several days until, while waiting for a tram to take her home one afternoon, her awareness of herself as a person disappeared. It was as if her head split open and her sense of self fell out. Following that she went through a ten-year ordeal of trying to live in a world and bring up her daughter without a sense of self.

This state, incomprehensible to me in my world, was a terrifying experience for her. Bernadette Roberts wrote three books on the self and what happens when all sense of it is lost. These major and challenging spiritual works serve as reminders that self-sense and intellect are not the same. In reading the work of these two extraordinary people, it is easy to see that neither Roberts nor Suzanne Segal suffered in any way from impaired intellect, though their logic and that of the common domain might not be quite in sync.

After a ten-year search for some clue as to what had happened and what could be done about it, Segal broke through the block of fear that had swamped her circuitry. Apparently it had been fear that had prevented her from making what we might call a final translation of her experience. After this breakthrough she experienced a fusion with what she could only call "the vastness," and entered into a reality that by her brief and simple description was as noble a sustained mystical state as any other recorded. Four years later, having sketched out her *Collision with the Infinite*, as she called her account of the venture, she died of a rapid-growth brain tumor. She was forty-two.

WHY TRANSLATION?
▼

The critical step in Laski's series—step six, translation—begs these two questions: Why must the answer be translated, and why is translation often so difficult? First, it's important to characterize such "answers" a bit more. They virtually always appear to their recipients in metaphoric or symbolic form— an imagery meaningful only to the recipients. Second, seldom does a recipient's answer bear any resemblance to the body of knowledge from which the quest arose, though it is to this background that the answer must be related for its translation. Sometimes the answer comes as a vast wave of knowing, like spiritual revelations whose contents must be carefully nursed and cajoled forth into some kind of verbal expression. Segal's experience brought some rearrangement within her brain itself and her translation was the long adjustment to that change, her discovery of its meaning as she went along with the change, rather than her attempts to reinstate her previous mind-set.

Translation is both necessary and usually difficult because the answer is discontinuous with the field of endeavor out of which the question arose. The answer is never made up of the materials gathered in the search for that answer and the gestation period never brings about a novel synthesis of current knowledge. Rather, the quest can give birth to something discontinuously new. Gould's analogy of "bricks and mortar" in the hopper of mind doesn't really work, because his answer was outside all knowledge within that field. Even though his answer could finally be translated only through using those bricks and mortar for its expression, something fundamentally new had taken place. Though quaternions may have been an existing potential in mathematics before Hamilton, and hexagons may have been present in chemical structure before Kekule, lasers are not present in nature itself even after Gould's revelation. Because of his *Eureka!* something that did not exist now does, or rather can, if you build the appropriate electronic device to bring it about. The same qualification might be made for Hamilton, or for Kekule, though this possibility is not considered by conventional academic thought and can be deeply offensive to our rational mind. There are times when the line distinguishing discovery from creation can't be drawn. That event which we assume is discovery may well be creation as well. Why, then, if the materials gathered have no relation to the answer given, is it necessary to undergo the long search for the answer? We are advised in the sayings of Jesus to knock and the door will open. If it doesn't, we must persist in our knocking, often to great lengths. Keep pounding, he advised, and eventually the opening must come, subject matter notwithstanding. Gordon Gould, however, unlike Kekule and certainly Hamilton, was not knowingly pounding on a door. His was a case of lightning striking on a clear summer day—and so was Suzanne Segal's. But as with Segal, Gould's bolt struck as a result of a buildup of resonant potential—a resonance that might be ever so subtle and a buildup that might be unknown to the targeted recipient until after the fact. But when all conditions are present, the intelligence of field function breaks through to an intellect of a like order. (A variation of this process lies behind the idiot savant phenomenon.) Bear in mind that matter is an aggregate of frequencies, neurons are aggregates of frequencies, and neural fields are resonant with fields of frequencies. Like attracts like.

Those who, like Gould and Kekule, receive such answers attest that the responses are not of their own making but arrive instead out of the blue, catching them by surprise when they break through. Such an answer comes through the neural circuitry of brain, but not from it—which explains why our mind must be clear in order for the answer to arrive, and why the circuitry must be in place ready for the answer, a readiness brought about by the passionate gathering of materials.

Current research shows that the brain is far more flexible than we thought, shifting and changing according to stimuli from the environment. At the risk of trivializing an awe-inspiring process, consider that the initial question, step one of Laski's six-step process prompting the search for materials, has its origin in the novelty impulse of the left hemisphere—prefrontal connection. This circuit is then sparked by the passion of the right hemisphere and its connections to the emotional-limbic brain, which is itself the connection to the heart. Through the orbito-frontal loop, the prefrontals connect to the limbic and heart as well as the neocortex. The heart's EMF functions holographically with all the fields of potential and draws on materials from them, the body of universal knowledge. When step five's answer forms, that energy follows or retraces the same neural route the seminal question and search originally followed, through the limbic-prefrontals into the right hemisphere. The holistic right hemisphere can't feed the material to the left in the neat, digital, linear style of left brain thinking, but only as that single whole knowing by which the right hemisphere functions. So at a point of critical mass, the answer arrives, unbidden, in the right hemisphere, and traverses the corpus callosum to the left, literally a discharge of energy from the creative to analytical structures of mind. This whole knowing appears to the left's digital process as a lightning bolt gestalt of metaphoric and symbolic form.

The corpus callosum can complete the circuitry only when the left hemisphere is inactive, when the analytical and critical processes of mind are suspended—you can't have reception of a whole gestalt and a dissective analysis functioning simultaneously. Given the answer, the interpreter mode, our translator, must then get to work dissecting and analyzing the intangible universal into its tangible variant. Note that the actual creation of the answer as itself is still unexplained here because it is still a mystery, as is the savant experience. We might say the answer forms within that hierarchy of fields in which our heart field is nested.

WHY LIGHTNING?

▼

Most recipients of the *Eureka!* intriguingly describe their experience as a lightning bolt. Real lightning, the kind in storms, strikes only a primed, fully charged target. Oddly and symbolically, the great mass of energy for a lightning bolt is built up in the clouds while a corresponding, opposite, but minor charge builds up in the ground. Both may be collections from a wide area—in fact, the ground charge may travel over quite a stretch of ground, running right through us, standing our hair on end as the charge searches out and collects its energy. Eventually, the two charges seem to seek out each other, and when they reach the closest proximity to each other, the weaker ground charge makes a gesture up toward the clouds through any medium that will offer it a bit of lift or passage, such as a tree. This invitation galvanizes the great cloud collection, which follows (chases down?) the weak trace until the two fuse in a wild rumble that brings a remarkable package of nutrients to the soil—each year many millions of tons of the soil's nitrogen are produced just this way. The earth has sowed a small wind and reaped a whirlwind from the skies. Or, we could say the earth asked a question and the sky gave its answer.

Creativity can involve a union of forces in both our creative passions and our quests through a process that functions the same regardless of the character or nature of the quest or passion. If you sow your small store of spiritual energy without reserve, brace yourself. Your answer, or lightning bolt, may have a corresponding cloud charge waiting in the wings. Bear in mind that only a neural system long immersed in the field related to the quest is capable of attracting and receiving the field's answer when it does form. Gould may have reaped the sowings of many like minds.

IF YOU ARE THIRSTY

▼

The *Eureka!* process is similar to Sat Prem's statement, "If you are thirsty, the river comes to you. If you are not thirsty, the river does not exist." Sat Prem's observation was made in relation to matters of the spirit, but if we pursue the issue long enough, we find these words as universal as spirit itself.

If our belief is passionate enough, the river comes to us and in whatever form the passionate belief makes possible. Belief is causative and passion is formative. Passionate belief is the chaotic attractor that lifts chaos into its particular order.

Not only does the person caught up in a quest generate the new potential, he or she also builds, through the rigors of the pursuit, the neural pathways that can then offer the avenue for the expression of that potential. Should some great revelation break in on my unprepared circuitry, it might well blow away my fragile wires. Thus creation's seeds sprout not on barren, unprepared soil, but on soil enriched, tilled, and ready.

The search for the materials, Laski's step two, isn't merely for the sake of accumulating them, as we always presume, but to build up appropriate neural circuits for receiving the answer. Without such opportunity for preparation, should seed fall in a haphazard way, such as Hamilton's quaternions falling in my head, no one would know about it, least of all me, for I would never recognize that seed for what it is, much less be able to bring it to fruition. The relation between neural circuitry and possibility is like the dividing line between being and nonbeing, thirst and river, creator and created. You can't have one without the other, nor both at the same time.

Mozart's description of the way his own creative process worked at times, in his mature period, offers a different facet of creation while reinforcing the field theory and the notion of readiness to receive the created. After receiving a commission for a piece of music, if he held the commission in his mind as an open-ended possibility, at some point the work would break into his mind as a "round volume of sound." A whole concerto, symphony, quartet, or sonata, complete down to every phrase, note, and nuance, flashed into his mind in its complete form, a unity of perfection. No matter its length, this perfect whole was perceived in a single instant.

It was an immense task to translate this instant's *Eureka!* gestalt into the thousands of individual little inkblots on paper so that an orchestra could translate it into our world of sound. Mozart, like the mathematician Stephen Hawking, might spend days working out the translation in his head. When this inner translation was complete and he was ready to turn it into the notes on paper, he would often have his wife read him stories to occupy his ordinary attention so that his musical mind could transcribe the thousands of notes more easily.

But to liken Mozart's creative procedure to *it* breathing through him, as though he were just a channel, an amanuensis for the muse, would be at best a disservice. His own comment was, "No one knows how hard I have to work at this," referring to the difficult translation stage before all was ready for the easier part of pen and ink.

Mozart had reached beyond the notes to find the music in a way few musicians do, exemplifying Blake's observation that "mechanical excellence is the vehicle of genius." At the instantaneous presentation of the "round volume of sound," Mozart had fused with the field of music as itself, rather as Segal had fused with the vastness. For the "answer" to move through Mozart, the neural pathways had to be formed and constantly tended, the soil tilled continually, even though such control must eventually be released to allow another part of the self to manifest—a perfect example of intellect and intelligence in balance, and an example of that gospel saying, "To him who has it is given more and more." To Mozart much was given because it could be received.

FIELDS OF INTELLIGENCE AND THE INTELLIGENCE OF FIELDS

▼

We should see that a field of information such as mathematics, music, optics, or physics is not some molecular-cellular memory collection but an intelligence itself or an aspect of intelligence, a potential capable of acting intelligently when the interactive conditions are right. Physicist Bohm spoke of consciousness "expressing itself" as energy or matter, to which we add intelligence. Earlier I observed that unity, whether an overarching universal or the particular unity of a field of knowledge, can't be known. Such knowing can be directly experienced only in mystical transcendence and might then linger as a kind of reflective wisdom, but this doesn't lend itself to verbal knowledge dispensed as that commodity of our day, information. Only unity's expression as diversity can be known. I select from a category in my computer a specific bit of information and this, while electronic, is essentially a mechanical function. But the mind and the universe it draws on and feeds into are organic, a living, pulsing process. The *Eureka!* answer, when it arrives in symbolic and metaphoric form, is a transition point between unity and diversity.

Consider that Gould's twenty-year pursuit of physics and optics fed into a unified field of like resonance. At a point of mass intensity in that field, Gould became an attractor sparking that field into spontaneous creation of an order it did not necessarily hold as a potential before. Perhaps he was simply in the right place at the right time, though such coincidences seem to happen only to a superbly trained mind, and though that mind must, at the moment of inception of an answer, be at rest—that is, doing nothing.[1] Any field of intelligence may spontaneously create within the nature of its par-

---

1. And here we have a tricky issue. It may not be the case that we can skip all the stages of doing to arrive at the point when that moment of not-doing pays off. Carlos Castaneda's form of not-doing required intense and unbelievably rigorous attention, risk, and a high level of doing before it could take place. Gordon Gould was correct that he had to feed the bricks and mortar for the new edifice into the hopper of his mind, though what emerged had no relation to the original bricks and mortar.

ticularity, but only in conjunction with a neural field of like order. This creation, we then say, pops into someone's head, but only into someone who can receive it. The creation has moved from its nontemporal, nonspatial nature or nonlocality into time. A neural field of brain and mind and field of potential have formed an interactive dynamic.

We may tend to assume some supercomputer in the clouds engineering all these maneuvers according to some master plan or whimsy. Instead, consider that intelligence unfolds in an infinitely wide diversity, each phenomenon unique unto itself. For instance, on reading the book *Microcosmos,* by Margulis and Sagan, I learned of mitochondria, one of the earliest organisms to appear on earth along with blue-green algae. Blue-green algae can convert sunlight into food while mitochondria can convert food into energy available and appropriate to cellular life. Nature has kept both invaluable functions intact, apparently unchanged, throughout the billion years of evolution since their first appearance. Life as we know it is absolutely dependent on both. Each cell of our body is loaded with these remarkable tiny living creatures called mitochondria, providing energy in the many forms needed.

Mitochondria have a unique DNA structure that is incomplete yet virtually immutable (hence their unchanged nature). But this incomplete DNA is completed by the DNA of the cell it occupies, and it is through this ingenious device that the tiny mitochondrion powerhouse can efficiently carry out the various energy requirements of any cell. For instance, when a male produces a sperm, the tiniest of all cells made by the human body, these even tinier mitochondria respond according to the unique needs of that tiny sperm. The sperm's remarkably long flagellum, or tail, by which it propels itself toward its goal, is made of nine microtubules (mysterious oscillators in themselves). A mitochondrion attaches itself to a gap juncture at the base of each of the microtubules, and from there it furnishes the energy to power each in perfect synchrony with the other eight mitochondria at their microtubule stations. This concerted effort begins only when the journey toward the ovum is actually under way, so that the generated power supply is not wasted. But how do they know when that time has come?

I read one research report that claimed that right behind the head of the sperm—which is but a ball of DNA, the payload the creature is designed to deliver—yet another mitochondrion sits. This one withholds its power play until the actual golden grail of the ovum is reached and breached, at which point the mitochondrion looses its thunderbolt in one glorious burst to boost the sperm's DNA into its promised land. (I understand this head-based mitochondrion observation has been challenged by other researchers, but the example fits too well to resist using it here, and it may yet prove valid.)

Now surely all this could be explained as chemical attraction-repulsion, if you like, as old-line scientism would. But so could my ability to read Margulis and Sagan, or my capacity to write this book, or yours to read it, which is a catchall cop-out that can be applied to everything to avoid any further explanation. Consider instead that what takes place with mitochondria is an intelligent action meeting a critical need of life, yet on a level so tiny that only a powerful microscope could detect it. Quadrillions of these mitochondria work around the clock in a vast array of tasks, to keep my body going. And I was aware of none of this astonishing procedure throughout my comparatively long life until I read Margulis and Sagan's study—and even so, I am aware of mitochondria only intellectually, an awareness that is but information. It does not give me the capacity to function in intelligent dynamic with that mitochondrion's intelligence. Our frequencies don't match.

Here then are two levels of intelligence, one microcosmic, the other macrocosmic, each completely unaware of the other. Were a mitochondrion to get ill or malfunction in some way, or should it not like its particular assignment, I daresay it would be futile for it to pray to some higher intelligence for assistance, for the highest intelligence in its universe would be me and I am completely cut off from it, at a radical discontinuity with it. Even after reading about it and its miraculous achievements, I as an intelligence, and mitochondrion as its own intelligent action, are still universes apart. Our discontinuity is almost complete, bridgeable only through artifact of thought or instrumentation. And the wonder is that neither of us needs to know of the other to function sufficiently.

From this model it follows that it is a bit naive for me to assume that some supercomputer master intelligence up there on cloud nine is orchestrating this infinitely diverse display of organic life, happily directing mitochondria onto each of nine microtubules on each of several hundred million sperm I (apparently still) produce daily, in just my small hide. It would be equally naive to assume that such gargantuan superintelligence is available to receive petitions from this particular mitochondrion called me. That a master intelligence running this show is also a personality, like ourselves, subject to the same influences we are and therefore on our same wavelength, may be equally naive, a variant of the old projection onto cloud nine.

Suzanne Segal, fusing with the vastness, exclaimed that "the vastness doesn't know anything is wrong." Nor would I know anything is wrong with my mitochondria, should they malfunction; I might only discover, perhaps, that I was dying more rapidly than usual. Intelligence is a vast mystery, but it operates on our discrete human level only as our discrete human intelligence, and even then we are aware of it only on our discrete personal level.

Yet Segal had fused with that vastness and, as was the case with Bernadette Roberts in her exquisite mystical experience in the Sierras, she perceived an immense and awesome intelligence permeating the universe. Blake claimed that a cup can't conceive beyond its own capaciousness, a classically logical observation. From this and the concept that like attracts like, we can surmise that Segal's and Roberts's cups, while ordinarily discontinuous with that vastness, were nevertheless, on some level, of the same order as that vastness perceived. Their intelligence was of the same nature as the intelligence on the continuum encompassing all intelligences and life. Theoretically, their intelligence could fuse with any microportion of the intelligence on this continuum, being of the same substance. Such fusion, however, does not invalidate the discontinuous nature of the functional expressions of that individual intelligence and intelligence as "itself," whatever that might be.

I thank God that I am not required by him or her or it to become aware of each mitochondrion within me and direct its function. Nor should I require of God such diversity, holding him or her or it responsible for every niggling action in an infinitely contingent process. There is a blessed gap between intelligences, or all would short-circuit in chaos. But there is also obviously a continuum within which infinite numbers of discrete actions can achieve a coherence and balance over and above any particular expression, as Mae Wan Ho described. Occasionally, when conditions are right, my small intelligence and that vast unity can resonate on the same frequency, which, though it nearly unhinges me, lets me know that *It* is there, wherever *there* is and whatever *It* might be.

Thus prayer on an individual level can be efficacious when all conditions are right, not by winning through worshipful flattery the favor from some superintellect on cloud nine, but by establishing resonance between discrete intelligences. Intelligence by its nature always moves for well-being when connection can be made. A prayer that leads to healing, then, might function much as a collective electrical charge that gathers to arc the gap to a smaller charge in its opposite polarity.

Such creator-created possibilities do not apply to Eckhart's "God beyond God." What characterizes "God beyond God"? we may ask, for which the only answer can be "Nothing." The only characteristic that applies is that *it* has none. God beyond God is beyond the creator-created dynamic. The radical discontinuity between universal and its specific variant, as between that vastness that "doesn't know anything is wrong" and Segal, was bridged in Segal, without her knowledge or any action on her part. "It does everything," as she claimed. The nature of both poles of the dynamic she experienced—the one universal, the other specific—were of the same order. And that order was, from our vantage point, love—the only avenue we have to the vastness, the only voice with which we can speak to that unknown, and the only response the vastness can make in turn. Nothing more, after all, is needed. But if we sit in our pain and anxiety, howling and weeping to that vastness to bail us out, the silence that meets us is not the silence of indifference—it is due to the simple fact that we have not spoken anything to be heard. The frequencies don't match. We seek magic when miracle is at every hand.

Gordon Gould's discovery was not simply an unearthing of one of nature's secrets, for that would imply that laser light existed all along, though we know that it does not exist in a nature without us. Even now we must build a machine to create it. Neither was Mozart's "Jupiter Symphony" there in the ether awaiting discovery, nor Bach's "Brandenburg Concertos." Life is a stochastic venture of creation, and spontaneity plays its part. The dynamic of Kekule and the field of chemistry probably brought the benzene ring into being. Biologists Maturana and Varela claim that the eyes see what the brain is doing even as the brain does according to what the eyes see. Perhaps Kekule gave us eyes to see in a new way. But the moment our seeing takes place, we are convinced we see that which was already there—lest we be held responsible for what we see! (That we can and must assume responsibility even for what we see and how we see it is a powerful and threatening proposal that accounts for why Blake came close to getting hanged for sedition, and why Jesus was.)

In my first book I proposed that no line could be drawn between scientific discovery and creation. Once a discovery is made, we can't determine to what extent our passionately pursuing mind may have entered as a determinant in what is discovered, which is an expression of the creator-created dynamic. Admittedly, such a mirroring relationship is heresy to our current religion of science, just as it was to medieval thought or to the mind-set of two millennia ago. But this question of the boundaries of mind and reality hangs in our limbo, unrequited.

Consider that the God conceived and experienced by Jesus and brought into our species awareness through him was, as an experience, in the same realm as Suzanne Segal's—an expanded form of Gould's laser, Hamilton's quaternions, or Mozart's symphony. Without Jesus there would have been no Father in Heaven appearing as a possibility in man's history. And this benevolent father as conceived by Jesus could no more have been conceived by Moses than neutrons or neutrinos could have been imagined by Galileo. In Galileo's time our universe of experience and thought wasn't of the expanded nature or stature to comprehend neutrons, nor was the cultural mind of Moses of such an expanse or stature to give rise to Jesus.

Man's mind is a mirror of a universe—which in turn mirrors man's mind to some indeterminable and unknowable extent. A God of love was Jesus' laserlike gift to the world. Without Jesus there would be no awareness of love of the kind that he offered. Jesus, then, brought his father into being inasmuch as God created Jesus. They created or gave rise to each other. This is why Blake said that to honor God is to love his gifts in great men—and why we should love the greatest men the most. Blake loved Jesus most of all men because Jesus displayed the greatest love and brought the possibility of it into our lives.

To experience Hamilton's quaternions we must build the necessary structures of knowledge involved, which involves a rigorous study of mathematics. In the same way, to access understanding of Gould's laser we must build the electronic machinery necessary to display it. To experience the benevolent father to whom Jesus referred, and thereby to know those hypothetical good and perfect gifts, we must allow our heart and spirit to build the neural machinery necessary to translate and display that state or activate the function that Jesus opened for us. We must be seized by the passionate question and pursue it with our whole heart and mind, as Hamilton did; we must keep pounding on that door, hanging on through the dark nights of those plateau periods. However you choose your metaphors, the creator-created function is the same, and through that dynamic we can make the necessary neural structures that will allow for such a God as Jesus' father to become a reality for us individually—and perhaps for all of us universally.

Jacob Boehme described it thus some four centuries ago, using the patriarchal metaphors of his time: The son will give birth to the father. We bring Jesus' father into our being and give his father being in the same dynamic. We can do this, however, only according to the model set forth for such a construction of mind—or live without such light.

TEN

▼

ALWAYS BECOMING

*To animate the world of the animate with objects, gods or geniuses is
closer to truth than with invisible deities.*

—WILLIAM BLAKE

Years ago, anthropologist Adolf Jensen gave us this classic report by an
Apinaye hunter from the Ge tribe of eastern Brazil. As you read it, bear in
mind Gordon Gould's lightning bolt on a clear day, described in chapter 9:

> I was hunting near the sources of the Botica Creek. All along the journey
> there I had been agitated and was constantly startled without knowing
> why. Suddenly I saw him standing under the drooping branches of a big
> steppe tree. He was standing there erect. His club was braced against the
> ground beside him, his hand he held on the hilt. He was tall and light-
> skinned, and his hair nearly descended to the ground behind him. His
> whole body was painted and on the outer side of his legs were broad red
> stripes. His eyes were exactly like two stars. He was very handsome.
>
> I recognized at once that it was he. Then I lost all my courage. My hair
> stood on end, and my knees were trembling. I put my gun aside, for I
> thought to myself that I should have to address him. But I could not utter
> a sound because he was looking at me unwaveringly. Then I lowered my
> head in order to get hold of myself and stood thus for a long time. When
> I had grown somewhat calmer, I raised my head. He was still standing
> and looking at me. Then I pulled myself together and walked several
> steps toward him, then I could not go any farther for my knees gave way.
> I again remained standing for a long time. Then I lowered my head and
> tried again to regain composure. When I raised my eyes again, he had

already turned away and was slowly walking through the steppes. Then I grew very sad.[1]

One of the gnostic gospels quotes Jesus: "I am always becoming as you have need of me to be." This is resonant with such comments as, "Before Abraham was, I am," or Augustine's statement four centuries later that there was never a time in human history when "that called the Christ was not present among us." Raimon Panikkar refers to the Asian Christ expressed in the Buddha, Krishna, and others. Whatever form it may take, this intelligence is always manifesting as the well-being of life, or trying to, as in the Apinaye's epiphany (*epiphany* meaning "an appearance of the god").

This becoming never repeats itself, for each situation is different and intelligence forms and manifests in response to a situation's needs. Such intelligence can move us beyond the constriction of culture if we can address him or her or it who becomes. Sooner or later each of us meets him in some guise. He invites our address, our acknowledgment of his invitation. Each of us gives way to fear in his or her own way when we are before him, as each of us knows sorrow when he turns away after we've declined his invitation. He is not the property of any institution or philosophy, he is not bound by religion's tradition. No man knows his comings or goings.

Had the Apinaye been able to make that response, I dare say he would have fused with him who beckoned. Our transformed hunter may have then presented his society with a new way of being, thereby lifting his people up by his presence. (His people may well have bashed in his head for his trouble, but he would have been compelled to make his gesture on their behalf.) Such transformation has probably occurred to lone individuals since time immemorial, as it did to Jesus in that desert. "If I be lifted up I draw all people toward me," he said—and one can be lifted up in many ways, and may need to be time after time. Indeed, Blake said he had died and been resurrected time and again in his life.

When God appeared to the hunter in the account above, that God was every bit as real and valid as the one who appeared to Jesus, Krishna, or the goatherds of Medjagorge. Each was the intelligence of life acting intelligently in a particular case—"Christ coming again"—according to the

---

1. Adolf E. Jensen, *Myth and Cult among Primitive Peoples* (Chicago: University of Chicago Press, 1963).

intelligence of those who asked the right question at the right time, or who were simply in the right place. The term *Marian phenomenon* refers to the current rash of appearances of the Virgin Mary worldwide. The one at Medjagorge, in Yugoslavia, mentioned in chapter 4, was the intelligence of life manifesting to several teenage Catholic goatherds, intelligence appearing not as the Buddha, or Krishna, nor as Durga on her tiger, but in a manner comprehensible to those youngsters. Similarly, intelligence would not appear to the Apinaye hunter as a radiant, white-robed being holding a shepherd's crook because that simply wouldn't be an intelligent act; it wouldn't register with the hunter. The Apinaye cognizes his own kind as we all do, and entered into the appearance of his God as the supreme Apinaye hunter, war club and all, much as Hamilton entered the field of mathematics, Mozart that of music, or Jesus that of his father. Creator and created enter into each other as the creation.

THE BOND OF HEART

▼

Jesus spoke of a narrow and a broad way to his kingdom, suggesting a kind of Darwinian selectivity that gives me a chill. A narrow way, however, admits only one person at a time—and one without baggage. The journey into the heart and the unknown isn't a group tour, as George Jaidar pointed out. And all this ecumenical business so popular on the cultural front today may not give the strength or clarity we expect for the journey any more than a merger of bankrupt firms gives solvency.

Our problem as an enculturated people lies in the combination of our lack of individuality and our isolation from our heart. Seekers of various goals gather in groups, thinking that through sheer number they will force the gates of wisdom, spirit, community. (Thus the fanatical proselytizing of religions. Enough members, and perhaps those irrational beliefs will become the case.) But group mind can't give community, no matter the numbers, and can only replicate its boundaried conditions. Community arises in any situation for that individual who has broken from group mind into the bond of heart.

Effective action is personal, not social or cultural. The minute our focus shifts to changing the behavior of society or culture, or any other person, we are projecting out and away from ourselves the solution of the way and moving toward tyranny. Two or three gathered in Jesus' name, for instance, is probably about maximum. More than that and someone begins to take charge, the cultural demonic sneaks back in, and we are soon at each other's throats, business as usual.

Jesus' unwritten word plunged us into a dialogue that can never be brought to closure. In this dialogue the real question is never what did Jesus really say, but what does he say to us, individually, at this time. His gospel can take many forms, as needed. Our mind's interpreter mode, whether personal, group, or cultural, having interpreted or translated Jesus' action into creed, wants to bring that issue to closure according to our interpretation. We each strive to speak the definitive word, just as the church fathers did, bringing others to the silence of acquiescence, as James Carse put it in his book *The Silence of God*. State-religion has tried to bring Jesus' open dialogue with humankind to closure, making the definitive statement after which nothing more need or should be said. The result has been a religion in which one need only believe the written formula frozen in print and interpreted by authority—an authority gained not only by killing a lot of heretics, but by killing the gospel as well.

We function holographically within a continuum of holographic process. Our individual well-being is just as important as the well-being of the whole because, in a holographic process, big and little, important and unimportant are fictions, games we play, boundaries we erect around ourselves. We are, as physical creatures, as big to the smallest atom as we are small to the biggest galaxy. The relations are the same. The very hairs on our head might well be numbered because we are, as individuals, as much what *It* is all about as anything or anyone else. Scientism, preaching its doctrine of despair, belittles our existence as insignificant, incidental, a minuscule accident of chance. Religion, preaching its doctrine of sin and salvation, belittles our existence as incidental and insignificant in the light of its supposed property: eternity.

Jesus described a higher evolutionary cycle that life is trying to bring about. He sought to wean us from archaic mind-sets that are no longer useful. He was hardly advocating a return to a hunter-gatherer mentality, as some have suggested, but rather a new mode of mind of which we are but dimly aware. This mode of mind, as HeartMath clearly demonstrates, functions as readily and easily in the middle of Main Street as in meditation or prayer, in forest or ashram. In the dynamic of creator and created, we are the mirror of a reality reflecting our actions back to us. Worshiping in spirit and truth is throwing caution to the winds and remembering who we are. In this dynamic, piety and treacly Bible talk is an offense to the intelligence of the heart, an unsubtle attempt at spiritual seduction. Whom are we trying to butter up with our simpering, "worshipful" words? What manner of god could be won over by unctuous flattery or flowery praise? Examine the doleful nature of most petitionary prayer or hymn texts. A maudlin sentiment saturates religion and is both a cloak for and an expression of self-pity and the feeling of being victimized. Sentiment and self-pity are alliances with death, light-years from a life lived in spirit and truth, open to the heart. Once open to the heart, we recognize the universe as benevolent and our personal self to be the center of that benevolence. The moment we place that center outside ourselves, as onto some group, person, or invisible deity, we have betrayed and denied our heart. The intelligence of the heart, as it moves for well-being, is not just a figure of speech; it is the only intelligent function.

HINDBRAIN BEFORE FOREBRAIN

▼

Jesus stated that if we will take no thought of tomorrow and "seek first the kingdom," all our needs will be met. (A simple, practical way to actually put our priorities in proper order and "seek first that kingdom" within, moment by moment in everyday life, will be explored later in this chapter.) Putting our needs up front places the cart before the horse, hindbrain before forebrain, and we are in stalemate. Our response has always been, "Once I take care of these basic needs, I will attend the higher," which action creates needs unending and shifts attention into the hindbrain. So we miss the best of both worlds: Needs are hard come by and the kingdom a fiction.

Our new mode of being has been given in our new neural structures of mind, and the model for activating those structures has been given as well. Our model is the one who could be falsely accused, tortured, and killed by his culture without losing his contact with the heart, without resorting to ancient survival systems, self-justification, or revenge. Breathed by "it," he allowed it to lead him to the greatest disaster rather than betray that inspiration and not be breathed by it. He did this as long as breath lasted, obedient to his heart, as was his ancestral Abraham, regardless of consequence.

And he did this as dictated by the intelligence of his heart. For through him intelligence moved for the well-being of our species as a whole, as it will with any of us today. Willing to break through our ancient defensive mind-set and reestablish heart's dynamic with intellect, we open that new field of potential in us. This can lift us up and out of our violence and suffering. Having seen such reunion lived out by our great model, we as a species should then know it can be done. The fire-walker, seized by the possibility of dominion over his own life because he has witnessed others walk the fire, takes the chance, risks his self, and comes into his own dominion, if only temporarily.

The analogy doesn't fit all the way for, insofar as the cross is concerned, we are not called on to reinvent the wheel—that is, repeat such a horrendous venture as did our model, or any variation. He advised us to agree quickly with our adversaries in this business, and quietly make our transition from culture to his way. Let the dead bury the dead can be read many ways.

IN SPIRIT AND IN TRUTH

▼

While resonant with the great prophets who preceded him, Jesus' answer was a leap beyond all their thought much as Gordon Gould's laser, we might say, was beyond Benjamin Franklin's kite. We can summarize his answer as love over law, making law obsolete and leading to Blake's insistence that you can't legislate morality or ethics, and that any attempt to do so is tyranny. The state of love Jesus referred to is a source of indeterminable strength, but is powerless and is attainable only by dropping a mind-set centered on power, intellectual prediction and control, and personal defense and self-service, all of which depend on or breed law in some guise.

Jesus didn't write anything because his *Eureka!* didn't lend itself to words but could only be lived. The worldview he pointed toward lay beyond intellect and logic. God had long been projected in many ways, projections that bred continual strife and endless theories about their nature. Jesus interiorized these centuries-old projections; he "owned" the ancient projections of God and realized his and our own nature in so doing. He proclaimed that God was not out there on some mountain or in some temple to be worshiped as a projected abstraction, but instead was here within us to be worshiped—if you must use the word—in spirit and truth.

Spirit is our élan, that "force that through the green fuse drives the flower," as Dylan Thomas writes. To worship in spirit is to give our vital energy wholly to the process of life, to hold nothing back, to make a total investment and risk of self—which means dropping defenses and giving ourselves to the intelligence of our heart. In return we are given far more than we risk. But this can happen only if we also worship in truth. The word Jesus used for truth was translated as *alethe* in Greek—*lethe* means "forgetting," *a-lethe* means "not to forget." What we are not to forget is who we are: cocreators with God, one with his creative force of life, not victims of it as we are enculturated to believe.

To not forget, or to live the truth, is to live every thought, word, and deed reflecting the kind of God with whom we want to be cocreators. A benevolent father giving good and perfect gifts was Jesus' version of our rightful heritage as the created end of the creative dynamic. To "do God's will" is to act at every moment as the mirror of that benevolence. Since this world is a dynamic of creator and created, the benevolence we live out will then mirror our action by default and the dynamic will grow in strength. To him who has it is given more and more. If we invested our meager coin in that market Jesus opened for us, our stock would grow. The field effect would take on more and more power for us. We would come into dominion over our world. Concerned for our souls, afraid of the loss of heaven and the threat of hell, moving into a defensive position, we lose everything.

The question arises: But what of those in group-mind who have no connection with the heart? As suggested with the Apinaye hunter, are they not likely to bash in the heads of those who do? Here, two cautions are given us: Don't let the right hand know what the left is doing, or agree quickly with the adversary. Our first and foremost task is to remain centered in our heart. Most adversaries only want their logical positions or semantic slogans of belief affirmed, and because no one changes their mind in conflict, with the old survival mode in command, the person in a state of forgiveness agrees on the surface (words are cheap) while remaining centered in the heart. We don't validate a belief or support it in doing this, but we do validate the erring person. We see God in our attacker or accuser and a bond forms beneath the surface. Our agreement disarms him and the heart has its chance to move and open. Cultural strife and contest are surface froth sustained by the energy given them. As Zen and Aikido show, displaying no contest in a situation while remaining centered in the heart can defuse that situation. We may even need to play the Sufi fool in this way of forgiveness. Ego investment, our image of self as standing strong for what is right and trying to defend or even force that rightness on others, verbally or physically, breeds conflict and blocks the heart. All rightness is intellectual and as such is provisional to a person who is one with his heart.

The overriding issue is that Jesus' answer to our current violence is diametrically opposed to our reigning mind-set. Our problems arise from our encultured minds but only a changed mind, a rearranged neural patterning, can understand why a mind-set can be the problem. A truly tone-deaf person doesn't understand what the fuss about music could be. We can understand the logic of Jesus' way only after giving ourselves over to it. Opening into a new mind and developing it comes from following that way. This gives the kind of knowing that can be known only through the doing itself, and only in the moment of that doing.

THE PATH TO FORGIVENESS

▼

A child in a state of unconditional acceptance of the given has no choice except to follow a model, for he has not yet formed the neural capacity for choice. By the time such capacity has formed, enculturation is complete and the child has no choice except to respond culturally. Jesus' challenge for us is to risk again that openness we had as little children—and it is a risk. We must become as we were before our defensiveness solidified the unyielding patterns sustaining culture.

The heart of Jesus' way itself is defenseless forgiveness, and here Blake's definition of forgiveness is clearest, as well as a dramatic reversal of pious platitudes. It is risky to present, however, because it can be used to cloak even the worst demonic impulse with sanctity. Blake's way seems poles apart from "agreeing quickly with the adversary." If we know what we are doing, though, Blake's pattern for forgiveness is valuable. Many injustices occur that are not adversarial to our person or our safety, and thus call for a different response.

The first step in forgiveness, Blake claims, is condemnation of sin. "Severity of judgment is a great virtue!" he wrote. Any and all restraints or hindrances of another are ruptures in relationship, or sins, and should be resented and denounced immediately and loudly, according to Blake. The weakly tolerant have no room in his world, nor, he noted, did they in Jesus'. Never is injustice or demonic action to be condoned.

But, as Jesus clearly demonstrated, the condemnation or judgment is directed at an action, not a person. All vengeance is evil, and just as the voice of honest indignation is the voice of God, resentment and retribution in any form are, in Blake's term, satanic. In our encultured mind resentment demands retribution and automatically seeks it. Northrop Frye points out that punishing a man strengthens culture, which strengthens mediocrity and reduces the human spirit.

The second step, then, in Blake's way of forgiveness is to separate the man from his error. But the only way open for us to separate the sinner from his sin is to make that separation in our own mind. Any attempt to change the other person is restraint of him and a mirroring variation of the wrong action. We judge the action, not the man, and then move to point out to the man the error of such an action.

This leads to the third step in Blake's forgiveness: the release of the imaginative power that makes possible this separation of the man from his action. That is, we must exercise our imaginative power to see God in the erring person, regardless of that person's action. Seeing God in the other is Muktananda's Siddha path and constitutes Blake's divine imagination. Jesus on the cross quite genuinely forgives those who put him there and in so doing illustrates how far we must be willing to go in forgiveness.

Northrop Frye points out that the prophet wants us to be delivered from evil, so he denounces the condition of the man who does evil. Culture wants only to be delivered from the inconveniences attached to evil, and so denounces the man by killing him or locking him away. The action runs rampant and prison populations double. Blake's prophetic vision would have us focus on a Messiah, the incarnation of God in the erring person, right there in the face of evil; the encultured mind would have us focus on a scapegoat. A perfect example of culture's "vision" is the "three strikes and you're out" legislation of the 1990s—legislation that imprisons for life the young man caught with drugs for the third time, while ignoring the conditions that created his addiction.

Seeing God in the other even as the other acts in an evil manner requires divine imagination, the ability to create an image not present to the sensory system. By observing a supposed objective truth out there, we become passive victims relieved of responsibility. Instead, we can use our eyes in active vision, to see *with*, not *through*. Rather than compounding our neighbor's error through intellectual condemnation of him, we can enter actively into that relationship through our heart. Freeing him from error in our mind's eye, we open him to deliverance in his own. There is only one heart, and it can take on any form.

Forgiveness is a state of mind, a way to live in the present moment, which means to allow each instant to pass without carrying negative elements of it over into the next. You can knock down your little child time and again and he or she will get up and come back, trusting and open-armed, time and again (at least until all openness is destroyed and defensiveness becomes permanent). We must backtrack, recapitulate, and realize we have no choice except to say yes to the heart. There is no judgment involved; it is a simple matter of frequency and sync.

In one of the gnostic gospels Jesus reportedly said: "Behold, I make all things new." Through forgiveness, intelligence can do the same, making things new moment by moment. Through leaving behind our history in which fire burned, we discover that it doesn't have to in our present moment.

We generally hold to a past event, however, whether it occurred seconds ago or years ago, in our desire to bring to justice someone who has offended us. We don't want our life or that of other people made new—we want revenge. Ironically, revenge is always cloaked as justice—and justice in this sense is always revenge. In this guise each present moment can only reflect the past and so we live in a hybrid of past resentment and future recrimination.

Gurumayi pointed out that the heart doesn't solve problems; it dissolves them and gives us a new situation in which we are freed from that problem. This is another aspect of forgiveness. It happens, however, if and only if we are willing to let go of the illusion of justice and righting wrongs. Such justice, while the lifeblood of culture, and thick in the blood of the enculturated, doesn't exist in the kingdom of the heart, where there is only just action, moving for well-being.

The Crucifixion and the Holocaust are the great injustices of history, and both reveal that there is no way of undoing an injustice or of justifying an unjust action. Note how acts of justice in our courts, international tribunals, or the arena of personal vendetta only serve to multiply the injustices they purport to address, often manyfold. That is why the early Christian act of putting the new wine of love and forgiveness into the old bottles of law and justice perpetuated an ancient and ongoing travesty of culture.

The Christian apology for this has always been to refer to love and justice as the two hands of God. But the gospel shows that we must choose between these two hands—that they, in fact, don't belong to the same God. The Lord of Moses says, "Vengeance is mine," while the father of Jesus offers forgiveness, love, and good and perfect gifts. Justice and love simply aren't the same. Justice cannot make way for love, nor can love make way for justice. Believing they can has led us into our perpetual war to end all war.

TEACHING OURSELVES HOW TO FORGIVE: HEARTMATH'S FREEZEFRAME

▼

After more than thirty years of research, HeartMath, a research center in Boulder Creek, California, has worked out a program of training that can bring heart and brain into synchrony. Its approach is biological, eminently practical, and entirely uncluttered with sentiment, and the synchrony achieved through its procedure can be as spiritual or mystical as we wish to make it, though HeartMath's most ardent customers seem to be corporate people who find that entrainment pays off on most mundane, monetary levels. Through its research and approach, HeartMath has lifted "thinking in the heart" from the confines of poetry and myth, which have not changed human nature, into the realm of biology, which can. HeartMath's program can free us from the survival defenses of our archaic brain structures and open us to higher heart frequencies. What we do once liberated is up to us.

Here is a simple six-step outline of the HeartMath "mind tool" called FreezeFrame. My explanation of the procedure is fairly lengthy, but the actual maneuver can be performed quickly and automatically once learned and practiced sufficiently. Although my translation is not a substitute for the center's training, it does throw more light on the heart-brain dialogue and, as you shall see, embodies in its steps the entire premise of this book—that we are indeed made to transcend.

1. Recognize when a stressful event is shaping up or taking place and "freeze the frame" at the instant of recognition. Freezing the frame is like pushing

the "pause" button on your VCR—the picture's action and sound are stopped immediately. As soon as you realize a stressful event is manifesting, freeze your state of mind, making no mental response. Any of us can suspend our thought, blank out inner chatter and ordinary reaction for a few seconds while we perform step two.

2. Shift your attention to the area of your heart. Focus and hold your attention there for the few seconds you will need for step three.

3. Recall a positive, joyful, fun-filled event in your life, or bring to mind some person whom you love fully or savor in memory. Form an image of that person or event as best you can and hold to the joyful feeling of that recollection without shifting your concentration from your heart area.

4. Keeping your focus on your heart, open to your intuition and common sense and, with utmost sincerity, ask your heart what would be the best response you could make to the situation at hand. What behavior on your part would be most effective in resolving the tensions or healing the rupture in the relationships involved in the situation taking place?

5. Listen to what you then hear or feel as your heart's response.

6. Act on the heart's response.

All this can take place in a pause between breaths, though it is slow going in the beginning and must be practiced assiduously. The biggest obstacle here is the deceptive simplicity of this description of the practice. We usually operate on the assumption that once we've heard or read about an exercise, we can perform it; the truth is that we need to practice and live with it for a while, or our associative thinking and rationale will dismiss the directions even as we read or hear about them. We might think, "Oh! I've done that all my life." And this keeps us stuck in our cultural mind-set, unable to open ourselves.

The rewards of the heart go far beyond simply reducing the stress of our life, vital as that is. But we must begin this practice on a small scale. Begin by using FreezeFrame on less important, incidental events rather than the major ones, which require more mental muscularity than we are likely to possess at first. As we discover the effectiveness of the procedure, we are ready to risk ourselves and take on larger challenges. As we become faithful in little things, we are made faithful for those that are greater. Finally, this small mental exercise can become a way of forgiveness, the way we meet life automatically.

THE SIX STEPS, IN DEPTH

▼

Step one is not simple. Ordinarily, the instant a stressful event forms we automatically identify with it and make what seems like a commonsense response. But caught up in the event ourselves, we are not merely perceiving it—we are that event. We identify with and accept it as naturally as we identify an itch we must scratch, which is just as nature designed.

Because of this automatic response, we usually recognize a stressful event after the fact—often long after. Whenever our inner state is agitated, we are reacting to a stressful event, whether of the present moment or somewhere in our past. With perseverance and practice, and admittedly some moral effort, we can shorten the time between the event and our awareness that the event was stressful. Finally, we learn intuitively to recognize the formation of a stressful event as it takes shape, and how to be aware of it without identifying with it.

If someone charges at us with fire in his eyes and fists clenched, our response is automatic and lightning fast. Such is our evolutionary nature— we protect ourselves. There certainly is no moral failure in this reaction. It is, in fact, our oldest evolutionary heritage. It is not, however, our newest evolutionary possibility—there is another response we can make, and while we developed the older one unconsciously, this new response must be developed deliberately.

We buy into a negative response so easily because our survival buttons are pushed, to some extent, by negativity of any sort and because a most complex cerebral loop generates from the button-pushing that is the fact of our life in that moment. All of this means we must have patience with our automatic reactions—they are more ancient than our species itself and we have lived them since birth. Simply reflecting back and remembering how a stress-inducing event began is a good start to completing the entire six-step practice.

To be able to step outside of ourselves, objectively observe that a stressful event is under way, and not react to it in the way we always have is a tremendous step. This is what evolution offers us, but, like providence, pro-vision (seeing ahead), or intuition, such faculties will develop only if modeled and nurtured by attention and practice.

Step two can be practiced mechanically. In Siddha meditation we practiced breathing into the heart—"Follow the breath stream to its lowest point and dwell there a bit," was Muktananda's way of putting it. Placing a hand on our heart region helps us to focus there. Bear in mind this action is not metaphoric in any way; it encourages a precise shift of attention and awareness away from our ordinary interpretive mind to that specific place in our body.

Concentrating on any arbitrary place or object won't work. FreezeFrame is not a diversion or sublimation, but rather a way to focus on the heart as a source of intelligence. It's important to be aware that in all probability we might have to pretend in the beginning, and might even feel embarrassed at our gullibility—but remember Shakespeare's observation that if we would possess a virtue, we must first assume it. Intent plays a critical role in this step; practicing it requires remaining intent on the heart and its intelligence to the exclusion of everything else. The first glimmering of positive results will dispel any initial embarrassment or feeling of gullibility.

Step three is what truly takes this entire practice to a new level. Steps one and two bring a shift of focus or attention that allows a break from our evolutionary past, which makes possible the actual shift of brain frequency in step three—something we can neither arbitrarily bring about nor think our way into. Here in this shift can be found the entire thesis of this book, for its moment is the moment of picking up the cross, of breaking the bonds of ancient instinctual behavior and thereby opening to the possibility of transcendence.

Ordinarily, when a saber-toothed tiger charges into our midst in any of its modern-day guises—an irate boss, a drunk driver careening toward you in his vehicle, an insulting and berating spouse, an angry customer—our ancient survival brains alert our high cortical structures to focus on the source of impending disaster. This lightning-fast reflex, millions of years old, shifts the intent of our highest forebrain's intellect into alignment with the hindbrain, compelling us to respond, "Look out!" In our ordinary state we have no more choice in our response than we do to something striking toward our eyes, which results in our automatic blink. In our ancient past, any creature who didn't attend such alerts left no progeny!

What we actually see with our eyes, a component of many of our instinctual reactions, is generated by our ancient animal brains. Our highest cortical system, where our human self-sense and thought seem to reside, then refines and interprets the imagery provided by our sensory system. By the time we see an external event, our lower brain systems will have put some forty million neural responses into action, as physician Keith Buzzell explains. These automatic responses include an instant report to the heart concerning the emotional nature of that sensory event. As a result, fast, reflexive intelligence goes to work, again as practiced by our ancestors for eons, which is why by the time we are aware of an event we have generally identified with it. This identification confirms the initial negative report sent to the heart by our lower systems. All parts of our system are then in synchronous action with our hindbrain to take care of the emergency. All the biological heart responses, outlined earlier, that were initiated by the lowest system have been reinforced by the higher structures of mind, which are now in resonance with the lower. All our forces have gone to war in the most natural way.

Ordinarily, there is no way we can be aware of this intricate neural maneuver because our conscious awareness is part of the dynamic. On our everyday working level, we are victims of our own brain-mind.

Such responses, however, are devolutionary for humans, who are given new neural structures for a dramatically different form of defense. These ancient reactive movements can be reversed, no matter how stubborn and nonnegotiable they may seem to be, so that new functions can unfold. At the very instant we perceive a possibly negative event, despite the fact that all the body processes alerted by the survival brain are being brought into play, we have our chance to act rather than just react, to intervene on behalf of our higher intelligence. Into this instant of chance, admittedly, a moral element enters, an opportunity for decision. We can act either reflexively or reflectively. Do we want cortisol overload and six hours of depressed immunity or peace of mind and harmony? Do we want retaliation or a new situation? Do we want forgiveness and freedom or justice and the prison of a consciousness filled with anger? Do we want an eye for an eye or a new way of seeing? And in the instant of this decision we cannot lie. If we think *yes* even as we give full rein to our reflexive reactions, our yes is pious but empty, with no substance beneath. This kind of yes means *no* will win.

In the first two steps of HeartMath's FreezeFrame, as we sense a negative situation forming, we freeze our frame of mind. We cannot stop the lower survival brain's automatic reflexes; those instincts have already gone into play by the time we are aware of an event in the first place! Because our high brain is the only part of the dynamic that we have any direct control over at this point, the one thing we can do is stop all high-brain reaction and carry out the action in step three.

A higher evolutionary structure of brain can modulate a lower system. The higher can incorporate the lower into its service as well as be incorporated by the lower. In stressful circumstances we have that cubic centimeter of chance to intervene—but we must first learn to recognize or at least be open to recognizing that instant of opportunity for decision and learn to make an honest response. Every event carries this decision factor within it.

Turning to the heart automatically serves the best interests of a situation as a whole, rather than the interests of ourselves alone, and little by little this begins to be the real benefit for us. This slow shift occurs not out of virtue but from a practical, larger perspective. We discover our true self-interests are met only when we focus on the total situation of which we are only one piece, and we learn that the intelligence of the heart can only function in this global way.

So the automatic reaction of our hindbrain is blocked by altogether opposite action: At the instant we recognize a stressful event forming, we shift our focus away from the impending disaster and direct it to the heart. In doing so we turn our back on millions of years of genetically encoded survival reflex and instead pick up the cross. In doing so, we accept on some level our death as an equally possible alternative within the infinitely open possibility the heart affords. Only in retrospect do we discover that the intelligence of the heart always moves for our well-being.

Such a decision must be made anew in each stressful event. There is no grand conversion to this intelligence of the heart, no theatrical emotional orgasmic fit that announces we are now realized and can relax. We must open ourselves again and again, for that instant of decision is ongoing and this new intelligence, rather than being encoded, forms only by our doing.

THE PROMISE OF REFLECTIVE ACTION
FROM THE HEART'S INTELLIGENCE

▼

We open ourselves to the heart, as outlined in steps one and two, only to use our high cortical structures in step three to re-create an event that is diametrically opposite to the event taking place out there, in our sensory-motor system. Our re-creation is a positive event of love, joy, happiness, fun, and exhilaration. We recall the feeling that event originally gave us and hold onto it as best we can—and we do this right in the face of saber tooth. Even as the ancient animal brains send the heart an emergency alert, the highest cortical system, operating from its more advanced evolutionary frequency, sends the heart exactly the opposite information, an image/state of love, peace, joy, fun, whatever. The real "offense" of the cross is the refusal to defend ourselves through our ancient instinctual pattern, for such an "il-logical" refusal disarms the whole cultural power. We pick up our cross instead of screaming for help. We opt for a higher intelligence and evolution leaps at the opening.

Again, the heart's intelligence is love, and love is a frequency that can synchronize only with its kind. The lower heart frequencies are synchronous with the animal brain's frequencies that move on ancient, well-established paths of fight-or-flight. The enlistment of the creative forebrain to recall an event of love or joy defuses the defensive circuit and the reactions that follow any survival maneuver. We re-create a memory of how some loving event felt rather than dominate our senses by a mighty act of will or pit our great virtue against nature. Forgiveness, after all, isn't warfare. It is a creative act.

Recalling an event of love or joy through creative imagination throws out a high-frequency bridge from the prefrontal cortex to the limbic-heart circuit. The heart automatically reciprocates on that same frequency, lifting us into a higher level of the creator-created dynamic, defusing defensive reactions already in motion, and opening an order of functioning not available to either intellect or imagination alone.

Step three stabilizes the heart's frequency spectrum and allows it to remain coherent; focus returns to its broad, global, generic readiness; the body relaxes on its own as we open to intelligence, that force that moves for well-being. Through this step we are able to opt for the excluded middle or gray area in our either-or logic. Higher frequencies modulate or moderate lower ones, enabling spontaneous healing, fire-walking, paranormal phenomena, a way out of disaster, and more—all of the possible outcomes of unconflicted behavior and dominion over ourselves and our world and a far cry from the domination of science and technology, or enslavement to archaic reactive instincts, or victimization by wrathful gods.

Jesus often used a miracle to illustrate a point or give an example of our new potential. But the next day many of his followers had forgotten his modeling of the way. "Don't you remember?" he would ask in exasperation. The crack in the egg seals automatically unless we are awake and alert and leap through it when it first opens. Sometimes we must attempt this over and over, until the crack appears more predictably and we learn to leap.

Step five of FreezeFrame is as critical as any other. We listen for the answer, which, of course, must take place through the neural channels designed for the purpose. The intelligence of the heart functions, Gurumayi pointed out, by changing brain function. The physicist David Bohm once said with genuine passion, as he gestured from his heart to his head, "When that realm of insight and intelligence leaps up, it can take out the dysfunction of mind and make it functional in an instant." "Behold, I make all things new," our great model proclaimed.

FreezeFrame allows the heart to change brain function, at which point we know what we should do. We perceive the answer and the intelligence of the situation manifests—and the more we practice, the more direct that answer becomes, until finally it is as automatic, instantaneous, sensible, and matter-of-fact as any other action of brain-mind.

Finally, step six: Act. We must act on the answer when we perceive it, without stopping to think, for should we delay—even if only to think about that answer—the ordinary processes of our survival mind take over and intellect, rationalization, and reason resume command. Our ordinary intellect will lead us not to doubt what we have heard or felt within us, but to erase it altogether, as though it had never been. Reason, or the rational, can instantly overwhelm intuition and render love powerless. If we are not thirsty, there is no river.

If we are thirsty, the river is there.

CROSSROADS OF DECISION

▼

Decision means "to cut off, sever from." We get from this literal definition our words *scissors, schism, circumcision, incision,* and *deciduous* (trees with leaves that are "cut off" seasonally). At each unfolding moment we have a cubic centimeter of chance to make the decision to opt for forgiveness, the movement that opens us to the heart. But this is a choice we must make anew each moment—it isn't a once-and-for-all decision. We choose the heart by cutting ourselves off from the myriad reasons for offense, revenge, or anger offered by our culture. Culture's clear logic and rationale, its vast tangles of legalistic reasons for why we have every right to be offended, arise automatically, moment by moment, out of the infinite contingencies of life itself. Jesus' execution was a rational act within his culture, while forgiveness was deemed the irrational. Forgiveness, however, is the a-rational, a refusal to buy into the world's logic and give in to offense through defense.

Culture applauds the man of principle who stands firm in righteous indignation and makes noble gestures for justice, for such actions keep culture's cycles spinning. Men of principle are often leaders who convince us to imprison, electrocute, crucify, or make war. Only forgiveness in this moment can break the demand for justice and the cycle of sorrow that follows, century after century. The heart has no principle except love. Love and forgiveness are equivalent and both are a state of mind.[2]

Gandhi knew something of forgiveness but adopted the maneuver as a political ploy to overthrow the British—through which ploy culture sustains itself. Jesus, however, did not use his understanding of forgiveness to try to overthrow Rome or even the Israelite lawyers and religious leaders

---

2. In our current culture, having a temper or cool rage is considered a sign of macho male strength and is increasingly being picked up by women. If you have no edge of fierceness, you are a wimp. As a corrective to this, Allan Schore points out: "[N]atural selection favors characteristics that will maximize an individual's contribution to the gene pool of succeeding generations. In humans this may entail not so much competitive and aggressive traits as an ability to enter into positive affective relationship with a member of the opposite sex." (See Schore, *Affect Regulation and the Origin of the Self,* 255.) Positive affective relationship rests on one's ability to forgive, to give over each moment to the rebirth of another. "Letting go and letting love" may be a corny, trite phrase, but it is also our salvation and surely that of our children.

whose practices were both hypocritical and exclusive. Instead he operated on a greater scale, that universal realm where each individual stands alone and naked before God in each moment. And something much larger than political advantage was at stake: The evolution of a species was rekindled by that cross.

If we examine Jesus' prescriptions for behavior given in all accounts of him, we will find they were designed to break us free from our hindbrain's survival modes. Dropping our defenses and self-serving maneuvers, our intellectual passion for prediction and control, frees us from our instinctual reflexes. Only then can we be lifted into the higher heart frequencies made possible through the prefrontals. Jesus had no social or cultural revolution in mind (the poor, he noted, "are always with us") but instead worked for the evolution or transcendence of the individual.

To those not ready or willing to free themselves from enculturation, however, his claims and directives are more than heresy and following them constitutes advocacy of an anarchy that undermines all systems of culture, including any notion of church or religious organization, his principal targets of criticism. George Fox, willing to spend most of his life in prison rather than betray his insight into the gospel and his guidance from the Paraclete, broke through the institutional stranglehold on spirit in his day. In our own day we had Peace Pilgrim, another of those rare expressions of the gospel. She was not invited to address international ecumenical convocations or church councils, not offered a Nobel Prize for peace or large sums for her autobiography. She wasn't even taken seriously enough to be arrested for vagrancy. And she was a vagrant—one with no personal history, no name (that anyone ever found), no Social Security number, no job and no money, no place to lay her head, and no clothes except those on her back. At an advanced age she displayed various psychic capacities, great energy, and buoyant enthusiasm and optimism, and she lifted up everyone who came into contact with her. She lived the Sermon on the Mount almost to the letter and made a difference in her world, one of the many Second Comings that occur. Had she been crucified by a madding crowd, she would no doubt have gone down in history. But that had already been done, I suppose, and, as befits our age, she died in an automobile accident, still nameless and unknown.

▼

ELEVEN

WHY BOTHER AND WHO CARES?

Everything possible to be believ'd is an image of truth.

—William Blake

A reader of an early version of this book asked, "What's in it for me? What would I gain from some hypothetical living from the heart?" This prediction-control reaction is typical of our cultural conditioning. That which lies beyond our current mind-set is subject to the same creator-created dynamic and paradox. A transcendent state forms underfoot as we step out into it—*being* comes into being and can be known only through our movement itself. Our stepping out in a transcendent move carries with it our intent and expectation, which may or may not enter into the nature of what forms underfoot. Transcendence is a movement into the unknown. Nevertheless, clues to the open-ended nature of this transcendent capacity are intriguing to consider. The following will explore forms of transcendence that are both concrete and abstract, material and ethereal, earthly and unearthly.

THE ABSOLUTELY OTHER

▼

Recall Bernadette Roberts's report of "breathing a divine air" for weeks and finding, on a return to her normal state, that the earth was a living hell by contrast. While I can't speak for Roberts's divine air, I do know that following that nearly ruinous fusion experience of my fortieth year, a serious rage against God arose in me. How can it be, I protested, that such a state is possible for humans, yet is so rarely experienced in life, which is by and large a vale of tears? What sort of God, I fumed, would rig up such a travesty? I know, of course, that God has not rigged the system against us. Rather, God is always coming to terms with life's problems, making the best effort to transcend them from the creator side of the dynamic. On the other hand, the vastness—that universal beyond all knowing of which Eckhart spoke, God beyond God—may not be a part of this equation at all. Although we seem to make connection with that vastness at times, this may be by random chance. In my particular fusion experience I had tapped into a frequency realm, or it had tapped into me, beyond comprehension or categorization. No creator-created dynamic was there, no polarities, none of Blake's "necessary contraries." One could not be in that state and function in our usual living state at the same time. Effort to get there may be a waste of time. Perhaps this led Eckhart to pray, "Oh God! Deliver me from God." There is that which is absolutely other to all we know and label, and for which we long with a longing that may never be assuaged. Perhaps, when Jesus on the cross cried out that his father had abandoned him, the father hadn't abandoned him at all, but instead Jesus was leaving the creator-created dynamic behind as he fused into that highest state beyond God. Perhaps it was from this highest fusion that the Paraclete sprang, that connecting link of spirit between what had been radical discontinuities up to that time.

Wherever transcendence itself might lead, then, is something only a transcended awareness would grasp, and then only according to that transcendent way. Bernadette Roberts's description of our endless "journey into God" seems quite apt, and I have been led to doubt that there could be anything I might ever do to comprehend, much less penetrate, that from which our glimpses of the highest state spring. We might be flooded by it on rare occasions, but that flooding is a one-way gratuity, not a dynamic, two-way dialogue. Our dynamics apply to the more material and concrete vistas that open.

For instance, recently, in my early-hour meditation, I slipped into a contemplation of the universe as a holographic torus and was flooded with a clear insight or perception of that toroid reality function, a nonmystical inner picture. (That is, it was an intellectual, matter-of-fact Aha!, not a rapturous revelation.) The torus was such a staggeringly complex yet simple arrangement that I was stunned. But then a thought flickered into my mind as brief as a fork of lightning, which would translate into the lengthy question: "But Lord, where were you before you manifested this torus?" Since the universe is by definition all there is, where was this other that gave rise to the universe or lay beyond it? (This was inadvertently an inverted echo of God's challenge to Job: "Where were you when I laid out the foundations and dimensions of this world?")

My hopelessly localized question was spontaneous and sincere. But in a response that was just as lightning-fast as my query, my skull, brain, mind, and sense of self imploded in what I perceived as an instant and total smashing of my physical head and consciousness. I was left with the ridiculous image of a pecan in the shell being crushed under an inconceivably powerful pressure. This action was as equally and mercifully brief in its duration as that flicker of question bringing it about, and, though painless, the crushing was anything but pleasant, leaving me with the notion that I had been given a glimpse of something so far beyond my limits of blood, bone, and brain that I couldn't even comprehend the scope of my question itself. In short, though the perception was perhaps a kind of answer, I had seriously short-circuited at its reception.

BONDING AND DOMINION

▼

A major argument of this book has been that transcendence, the ability to rise and go beyond limitation and restraint, is our biological birthright, built into us genetically and blocked by enculturation. Were we to conceive, deliver, and bring up our young within the bonds of love, where our young would feel unconditionally wanted and accepted and were never betrayed by their matrix world, our full human nature might unfold with no more struggle than any other aspect of our growth. We do not have to struggle mightily to encourage or force those molars to break through at age six and twelve, or wisdom teeth at eighteen. The word *God* might never have been coined were we free-flowing expressions of God's creation, much as the word *healthy* would never have been invented were we never unhealthy. "Man is born like a garden fully planted and sown," Blake claimed. "This world is too poor to produce one seed." But we as individuals and our world as a whole must nurture and protect the seed we bring.

This is why Jesus made his aforementioned comment that to "cause one of these little ones to stumble" was a major, nearly irreparable crime.

And it is one reason at least that Jesus didn't refer to spiritual paths but to a way of being that opens only in this moment, for which there is no preparation, and that has no conclusion. Today is the day and this is the hour—moment by moment. In Jesus' way, an opening in this moment does not freeze the next moment into its likeness; each moment is its own creation. And freedom from concern over safety in the next moment can open us to the heart's providence in this one. (The word *providence* comes from *providere*, "to see ahead.") But we share that provident vision only as we open to it in its own moment of now.

Jesus' way, then, is its own goal and leads nowhere; it only exists as created anew moment by moment. Thus his way offers no place to lay our head, no final goal or stopping point, only a journey into God. In our shamed state of enculturation we feel we must be scourged and purified before we are worthy even to undertake a spiritual "path" and often demand that such a path be difficult and painful, leading perhaps someday to a place of rest and surcease of pain. Of all the great beings, other than Lao-tzu, perhaps, Jesus broke from this encultured notion of guilt and punishment, clearly pointing up the creator-created dynamic and our freedom from guilt or blame.

But we, strangely, can't let his way be in that simplistic state. We insist on creating an unending, ever-new supply of macho and tough spiritual paths with mountainous obstacles, urging each other to be brave, carry on, and not stop halfway up. We rig up graded systems of spiritual success to measure each other by, determining how high up the ladder we ourselves are, forming hierarchies of the superior professional athletes of the obstacle course, and hiring them to travel the ever-burgeoning lecture circuit. We seem oddly offended that transcendence should be our nature rather than our reward for overcoming nature.

All his way asks of us is that we invest our life, not hoard it; risk ourselves rather than waste our energy defending against the stochastic nature of this venture. The timid recluse afraid to invest his talent loses it. Apprehensive and fearful of the evils of life, the brooding contemplative retreats into his safe mental cave of reflections. Afraid to risk relationship, the armored person further arms himself in celibacy for "spiritual reasons." Worried about losing his soul, he bargains with God, fate, or destiny, trading the juice of life for a supposed safe space on cloud nine or a reservation in that house of many mansions. Such a timid, protective person isn't much a partner in a creator-created dynamic. To worship in spirit and truth is to throw our life without reserve into that "force that through the green fuse drives the flower." Jesus cursed the barren fig tree, not the Roman conqueror.

DOMINION AND UNKNOWING

▼

As we move beyond our survival mechanisms, we experience an increased dominion over our world. Dominion is quite different from domination; we try to dominate a world or nature that we feel victimizes us. Dominion comes from discovering that we can transcend limitations as they arise—thus we need take "no thought of the morrow." Mircea Eliade, the anthropologist and a follower of Carl Jung's psychology, lived for some ten years with Tibetan yogis early in the twentieth century and wrote extensively of the dominion the developed yogi had over his body and interactions with his world. He referred to their ability to "intervene in the ontological constructs of the universe"—their ability to function outside the cause-effect of ordinary reality. Eliade explained that where the faith is simple, the test of faith is simple. So that all may "know them by their fruits," the yogis must physically manifest according to their tradition, in their immediate body and life.

To illustrate his point, Eliade describes one exercise for graduation to the level of full-fledged yogi, which involved drying a stack of frozen, water-soaked sheets through the exercise of *tuomo,* or production of body heat, while sitting stark naked on a frozen lake in temperatures about 30 degrees below zero, a job that generally took all night for the novitiate to complete. This required a form of unconflicted behavior that involves our highest neocortex modulating our lowest sensory-motor brain. Through the same unconflicted behavior yogis could be immune to fire, although they never practiced fire immunity to the extent the fire-walkers of Ceylon (now Sri Lanka) did long ago and were still doing back in the 1950s.

Consider, however, that children who were conceived and brought up with an open frame of reference and educated to play with and expand the *boundaries of mind* in this yogic sense, rather than being governed by such boundaries, might not intervene in the mechanics of our world in this manner. The freedom to play with and expand boundaries rather than being limited by them was articulated by James Carse in his book *Finite and Infinite*

Games. It may be that we in the West have secretly longed for such magical displays as fire-walking and tuomo to prove the lie to our enculturated conviction of limitation, constraint, and being victimized, and, in our attempt to dominate nature in turn, we lay waste to her at every hand. Rather than manipulating or destroying nature, we can use her as the base from which our creativity and dominion can function. Some aboriginal populations did just this for many generations, even millennia, leaving little trace behind them. Michael Murphy, in his book *The Future of the Body,* lists hundreds of examples of developed as well as random cases of individuals coming into some dominion over their personal, private world in a manner similar to the yogi's.

In Jesus' sense of dominion, once we have been faithful in small works by which we learn to trust the life process, larger and larger works become possible. Where that could lead we don't know. In every case of my unconflicted behavior when I was twenty-three, the various witnesses to the episodes soon phased those events out of their memories and within a matter of weeks refused to discuss or acknowledge what they had witnessed. Memory can be selective on behalf of that which we think protects us against a "collapse into chaos should our ideation fail."

Just as the Tibetans practiced a form of unconflicted behavior, select groups of our species may have moved into and out of such ability time and again in history, discovering and then forgetting what they had found. And just as I remembered and forgot myself again and again, my friends forgot or screened out the crack in the egg offered through such experiences. The world was too much with us all, soon and late. But this may not have always been the case, as the following examples of dominion, and possibly transcendent states, can show. Consider the following anomalies, which contemporary academic thought screens out, as examples of unconflicted behavior on a broad, social scale, indicating societies that may well have gone beyond ordinary cause-effect and its limitations and constraints.

UNCONFLICTED BEHAVIOR ON A SOCIETAL SCALE

▼

Architectural remains in Baalbek, in the mountains of Lebanon, show that this site (named as the home of Baal, our Old Testament Yahweh's rival) was occupied by waves of civilizations for many millennia, reaching back long before the Minoans, each of them building temples on the ruins of those that had come before. Beneath the ruins from civilizations of which we have some record lies an immense stone structure of staggering proportion from a civilization entirely unknown to us. The lowest level of this structure consists of a base layer of stones, each measuring thirty-three feet in length by twelve feet in height by fourteen feet in width. The stones in a second course measure sixty-five feet by twelve feet by fourteen feet, with each one standing a bit shorter in length than the one next to it. Five of these stones laid end to end would cover the length of a football field, yet they are cut and fitted so perfectly that at first the structure was thought to have been carved from a single immense rock bed. Each stone has a single, small circular hole drilled through its exact center. The interior of this hole is larger than the exterior.

In the second course there is a gap of some sixty-six feet filled with thousands of smaller stones dating to a time apparently long after the building of this layer. The quarry from which the stones were cut lies about five minutes away by car over rough terrain. There the clear outlines of where the stones once were are visible in the mountainside, and the missing sixty-six-foot stone is still there, cut completely on three of its four sides. One massive seventy-foot stone lies on the ground, cut free and apparently intended for a third course. The progressively larger size of the stones may indicate corresponding progress in the technique developed to cut and move them; ordinarily building starts with the largest foundation stones first, followed by those that are smaller. This building project was possibly experimental and obviously interrupted or abandoned in midcourse.

Other than the hole in each stone, no clue to a possible technology for moving them has been found. British engineers estimated the weight of the smaller stones at 750 tons each and the larger ones in the second course at more than a thousand tons each. By way of comparison, the Egyptians worked with stone weighing two to three tons each, on average. The engineers estimated it would take the power of forty thousand adult men to move one of the larger stones, provided you could arrange to have that mass of energy from so many people organized around a single object, and a roadbed sufficient to support it. But there isn't a trace of a roadbed between the quarry and the building site.

Archaeologists claim that this part of Baalbek is antediluvian (before the Great Flood) and estimate that the structure was begun anywhere up to twenty thousand years ago, loosely within the estimated time frame of the carving of the Sphinx, a monolith of similar scale. The Old Testament speaks of "giants abroad in that day," and there could well have been—not necessarily giants of body but of mind and spirit.[1]

Recent archaeological studies of Machu Picchu in Peru, a structure far older than the Inca civilization decimated by the Spaniards and built of sixty- to seventy-ton stones, show that the top of that steep-sided pinnacle had been sheared off and the debris shoved over the edge of the mountain. The quarry used as the source of stone for this complex was recently discovered some five air miles distant. The quarry was actually formed by shearing off the top of another peak with near-vertical sides, and clearly shows the outlines of the various rhombus-shaped stones found at Machu Picchu. The terrain between the two pinnacles is jungle, with rivers, chasms, and terrain typical of the Andes. No sign of a roadbed exists, even if one could have been built there.

An archaeologist from the University of California at Santa Barbara who was involved in the recent research of the site pointed out to me that we have no technology today to move sixty-five- or seventy-ton stones from the pinnacle of the quarry to Machu Picchu. The only approach to either pinnacle is a single narrow footpath winding to the summit in a series of hairpin turns.

On the east coast of South America there is a one-hundred-mile stretch of coastline with a series of parallel, boxlike stone jetties running out into the Atlantic, all equidistant from each other, some hundred feet in length, and sixty feet high (if my memory holds). I'm not aware that any of these structures has been dated—perhaps they can't be—or that any reason for them has yet been dreamed up.

Here in North America, the mound builders appeared before the Christian era and disappeared an estimated three hundred years before the time of Columbus. Some of their mounds contained burial sites, but most did not. They served purposes of which we know nothing but conjecture much. Skeletal remains that have been unearthed are of men who average seven feet in height and women who average six feet in height, both with large,

---

1. See Michel M. Alouf, *History of Baalbek*, 15th ed. (Beirut: American Press, 1938).

round heads. The mounds in what are now Ohio, Indiana, and Illinois are of various geometric shapes, some clustering to cover large distances, and are obviously related parts of a pattern. The large mounds, most of which are many hundreds of feet in diameter and of perfect shape, were discovered by aerial photography in the 1930s and '40s. The famous serpentine mound in Ohio, magnificently precise in its execution, was estimated to have been originally some three thousand feet long.

The largest mound, found at Poverty Point, Louisiana, consists of six concentric octagons cleverly arranged with the points left open to form avenues, each of which points to some significant astronomical fixture, such as the polestar, the Dog Star at summer solstice, and so forth. The mounds are estimated to have been about sixty feet high originally, as were the others on the continent. The total structure at Poverty Point is three quarters of a mile across. Altogether, it represents one of the most massive movements of earth in history.

It is noteworthy that the American Indians who were here when the Europeans arrived have no physical similarities with the mound builders. Nor have the mound builders any apparent connection to the civilizations of Central or South America. No one has any notion of what happened to those people or any notion of what the significance of the massive mounds might have been. It is clear that such perfect geometric figures built on such a scale indicate an advanced knowledge of surveying as well as of astronomy and geometric design.

Archaeologists from the University of Louisiana, examining the ruins of Poverty Point, surmised that because the structure seemed astrologically aligned it must have been a kind of geometric calendar by which those early hunter-gatherers (by academic consensus, the only kind of people allowed on earth at that time) could foretell the seasons. Why would they want to do that in a semitropical world? To know when to plant their crops, which would indicate that they were in transition from hunter-gatherers to farmers. The need to obtain such seasonal information ahead of time in that lush area impelled them to undertake one of history's most massive movements of earth to create remarkable geometric monoliths.

While I, for one, thought the academic pundits who came up with this explanation needed to do a bit of hunting and gathering of common sense, regardless of explanation, method, or technology, it's clear that great civilizations displayed unique capacities we can duplicate at best only through complex technologies, if at all. Our response to such anomalies is to explain them away if we're scientifically oriented, or project them onto fantasies if our understanding leans more to the occult. A favorite modern projection is of gods who come down from the stars to give our poor species a boost. But we don't need to resort to gods from cloud nine to explain the seemingly impossible creations such as Baalbek, Machu Picchu, and the mounds of North America. They are evidence of a developed form of Piaget's concrete operational thinking, which a child begins to employ at about age seven and which is poorly developed in most of us. It involves the ability to operate on or change some or all characteristics of a physical process or material through the use of an abstract idea. If its definition is extended, it indicates an aspect of mind over matter, which is a final heresy to Newtonian-Cartesian thought. In the last analysis, concrete operations consist of the higher structures of mind operating on the lowest sensory-motor structures—as occurs in fire-walking. Nothing new here.

Concrete operations of mind may have been commonplace in earlier periods of our history following the full flowering of the prefrontal lobes, until, as hypothesized earlier, some sort of calamity shocked the species into a defensive recoil from which it hasn't recovered. Because of the self-replicating imprint a fear-based culture brings about in each new generation, we haven't yet regained our balance as a species on any functional, widespread level. We essentially lost our nerve and haven't regained it. The few types of operational thinking we have developed are products of a defensive intellectual orientation that makes our inventiveness dangerous and self-defeating in ways that may well bring about our extinction. There is nothing intelligent about a neutron bomb, germ or chemical warfare, atomic reactors and the production of plutonium or, in fact, many of our contemporary methods for species suicide.

I keep wondering about those civilizations—the one at Baalbek, the one that made Machu Picchu—that had some interesting tricks up their sleeves but apparently just pulled up stakes and left, or were wiped out. Perhaps, in an escalation of ability, they rapidly grew beyond playing with stone blocks and moved onto something else—formal operations rather than concrete, for instance, something we have not yet fully explored. Recall that the formal operations stage, according to Piaget, is when the mind matures and can stand outside of its own process and operate on it, changing the very function of brain-mind. Muktananda disparaged our notions of progress and superiority. He claimed that many great civilizations have risen and fallen, appeared and disappeared, over the millennia. Carlos Castaneda implied as much, and anthropologist Leslie White spoke of great civilizations rising and falling in successive waves and always falling by their own hand. It's clear to see in our history that seldom is a civilization aware that it is falling, and we are no exception, being too busy bringing about our own unwitting demise.

DEATH AND RESURRECTION
▼

The objective of this book is not to try to define or describe a transcendent state, but rather to present the biological truth of transcendence, which, I claim, was the central theme of Jesus' good news. As have all of history's great beings, Jesus tried to wean us from the limitations of our fear-based mind-set. The miracles of mind over matter were part of his ploy, his attempts to shake us out of our sleep, which is how miracles have been used by teachers in the East. But culture's hold on us and the violence among us that results are as much a concern today as they were in Jesus' time, if not more. No dramatic or romantic philosophy of "onward and upward to the stars" is going to help us if we don't come to grips with our murderous behavior toward ourselves, our children, and our earth. Long before we could bring a populace around to some New Age supramind, we will have done ourselves in quite thoroughly.

Our first task is to stop projecting onto romantic myths and assume responsibility for our part in the creative dynamic, starting with ourselves and our infants and children. With our present knowledge of brain-heart interaction, conception, pregnancy, childbirth, and child development we could bring about the most immediate and dramatic revolution of our history. Before we can accomplish this, however, we must get out of the defensive postures that keep us in servitude to our hindbrain.

Consider studies of cellular longevity: A cell from a rat, placed in vitro, will divide and reproduce five to seven times before it ceases to reproduce further and dies. Rats live some two to three years. Cells from more advanced mammals such as chimpanzees will reproduce from fifteen to twenty times, and these creatures can live as long as fifty years. Cells of a human will reproduce from fifty-five to sixty times. Thus we should, by the same expanding ratio, live a minimum of 150 years. Perhaps our bodily immortality is not what we should be after, but rather we should be addressing our premature morbidity and serious underdevelopment because of a foreshortened life.

Numerous studies show that eradicating anxiety and stress and their accompanying cortisol would, in itself, greatly increase longevity and decrease illness. Perhaps our rage is not so much over death as over getting cut off in midstream. I feel I have just begun to catch on to some of the nonsense around me, just begun to wake up even as my particular cellular system, long past its three score and ten, is falling apart. Perhaps my body is simply following the dictates of an ancient cellular memory carried beyond its time, as the late Indian sages Aurobindo and the Mother suggested. Even so, the notion of an immortal cellular body might not be so attractive to us if we had a good, full 150 years to mature and get ready for some new world to create and explore, perhaps in a resurrected body.

Above all, stress and anxiety might be largely alleviated if we realized that Jesus' good news was the truth. If we could accept that we are children of a good and loving creator, like sons and daughters apprenticed to their fathers and mothers; that we are what we, in our heart, believe ourselves to be; and that greater gifts of spirit than those displayed by Jesus are available to us, we could live in spirit and truth. We would abandon ourselves to our heart and allow our intellect to serve that intelligence. Were we bonded to our heart from our beginning and given models for its development, we would grow in perfection. As Blake and the Sufi claimed, what lay before us might then be limited only by our capacity to imagine.

Finally, in considering what kind of reality a full awareness of a transcendent state would bring, consider the notion of resurrection, an idea developed by the Egyptians and a matter of hot debate in Jesus' own time. Or consider simply that death of the body is not necessarily death of personal being, a concept compatible with Mae Wan Ho's liquid crystalline organism. After all, death is the ultimate limitation or constraint and a natural case for transcendence. Survival of personal being beyond death of the body is accepted as a possibility by the Sufis and discussed by Ibn Arabi. Surely something of this nature played a part in Jesus' great gamble. Indications seem strong that he thought he could pull off the great trick of the ages, come through the death experience intact, and so prove the viability or durability of a person's core being. This alone could break us out of our self-replicating cultural trap and would be the culmination of his whole magnificent gesture.

There are many aspects to this concept of resurrection, and a host of theories beyond our need here. I would like to suggest that the notion of an immortal self, mind or soul existing beyond death, may have occurred to the earliest true human on the emergence of the prefrontals. Through millennia of ancestral longing for immortality, we may have posited that possibility within the creator-created dynamic, similar to the notion of atoms being posited by the Greeks and brought to realization by modern science. This longing to survive death may have had no historical precedents before the greater prefrontals formed because without the prefrontals we would not be capable of imagining beyond our limits of body and brain. The possibility, once injected into the creator-created dynamic, may well have become a field effect, a projection that our creative mind fed into the hopper of creation, so to speak, like Hamilton and his quaternions, and requiring long periods of gestation and growth. Jesus, in owning the Hebraic projection of God and realizing it as a God of love, may also have owned the long-simmering projection of resurrection and given it some sort of close approximation. Until his time, the Jews had no concept of an afterlife.

It seems to me his magnificent venture into death worked on some level, and, though grossly misinterpreted, brought a warp in historical time we have yet to comprehend. We have been so engrossed in trying to make our projection of a cloud nine Jesus work that we may have missed the rich vein of gold he opened to us.

Like the fire-walker who models an impossible action others can then emulate, Jesus' feat should have broken the stranglehold death has on the human psyche. Fear of death locks our mind into survival strategies that counter our discovery of possibilities other than death. The biological cosmology of transformed cells proposed by Aurobindo and the Mother may have been echoes of this very feat of Jesus. The Mother and Aurobindo, and their disciple Sat Prem, felt that one person breaking through into the new modality would open the way for all, that shedding the conviction that death is inevitable would break the hold of that notion and its inevitable results. The history of Jesus as a model of this greater life shows, however, that a one-man shot can't easily dislodge the cultural stranglehold.

Being the exemplar of such an evolutionary breakthrough may have been a major impetus in Jesus' passion, and he may have counted on us buying into his field through the impact of his example. Because any field grows with usage, his kingdom would thus come about, as natural a process as our current enculturation. "To those who have, more will be given," he declared, but first we must garner enough of that gratuity to attract more. We easily do this with the demonic—why should the divine be so much more difficult to generate?

As pointed out earlier, among the many factors preventing Jesus' efforts from working in any direct manner is the mythical projection brought about by Paul's christological overlay, which then colored all further reports concerning Jesus. Paul turned this greatest of all human ventures—Jesus' willingness to die for his people, and the possibilities of his resurrection—from an evolutionary breakthrough into a tired Greek *deus ex machina*, a God-in-the-box trick upon which a religion could be built. In doing so, Paul followed an ancient reaction posture, and though he didn't invent it, the deadening effect is the same.

OF FIRE, FLESH, AND FAITH

▼

At each fire-walking ceremony in Sri Lanka an average of one hundred people walk into a twenty-foot recessed pit of white-hot coals that will melt aluminum on contact. True, an average of 3 percent of the participants die in such pits yearly, but the rest go through unscathed, ecstatic, and transformed. I have given details on this in several books, notably *The Crack in the Cosmic Egg*, and the phenomenon has been rigorously investigated, filmed, and scientifically monitored. The latest debunkers focus on the thin little beds of scattered coals in America's New Age pastime and make a fuss over thermal mass, but they ignore all fire-walking phenomena that won't fit their explanations.

The history behind these Eastern enigmas reveals the creator-created dynamic and is a cultural variation of Margharita Laski's *Eureka!* formation. For many millennia, yearly sacrifices to various fertility gods and goddesses were made throughout India and the Orient. A victim was chosen and declared to be a god's temporary incarnation; was honored, worshiped, and given every luxury for a year, including choice of women eager to comply; and was then duly anointed and sacrificed in various ways on the designated day. Sacrifice, in this case, is sacred murder. (Although the word *sacrifice* comes from the word *sacred*, meaning "whole" or "to make whole," its connotations have surely changed.)

In one area of India, when the great day arrived, huge hooks attached to long ropes were run through the back of the person to be sacrificed, after which he was pulled by the hooks to the top of a tall pole on a bullock cart. All day he was swung over the various fields to be planted, his blood watering the soil. On one momentous occasion centuries ago, according to folklore, the victim went into an ecstatic rapture as the great hooks were rammed in, and throughout the day chanted in ecstasy that he was the god as he swung, hale and hearty. At day's end, when the giant hooks were removed, they left no marks on him—there was no blood, no pain. Since that long-ago day, no injury from the maneuver has resulted, and the position is sought with zeal by those who would be gods. This ceremony was still taking place in India in the 1950s and has been both filmed and recorded.

In Ceylon (Sri Lanka) the representative of the god was thrown into a fiery pit as propitiation to the unseen deity. At some remote time that vicarious god went into an ecstatic state when facing his fiery death and walked through, exalted and unharmed. He had become, in effect, the god he represented, much as Jesus owned and became the long-projected Hebraic God and transformed it into a God of love. From that day on, the position was avidly sought by the Ceylonese. The god, Kataragama, became an amiable, benevolent one and his worship through fire-walking became a sacred cultural practice in Ceylon, the way of ensuring the god's benevolence for another year—and, as is true for all cultures that look above for benevolence, the way to continue to project a human potential onto cloud nine rather than own and become it.

Today, giving up life, accepting their death, the fire-walkers in Sri Lanka triumph over both fire and death—though they then give away their dominion by projecting their triumph onto their invisible god, giving him all the credit. Because nothing can then happen except through the grace of Kataragama, this limitation also becomes real. The dynamic of creator-created is not state-specific or selective; it is universal.

Field effects, once set in motion, tend to sustain themselves. Once established, such powers are then effectively outside and beyond us, available to us only through supplication and obedience to that field effect—or god. Nothing has changed. Religious faith today is an extension of that same primitive belief in magic and the same misplacement of dominion. But Jesus' Paraclete is still here urging us to possess the projection, own it, and become it. Greater gifts than his would be ours because he has merged with his field and become one with his father and so with us. But we would have to buy into Jesus' field, not a cultural counterfeit of it.

BODIES WITHIN BODIES

▼

Our cultural anxiety over our body's fragility is part of the air we breathe and not an easy barrier to break through, no matter how many glimpses we get of something beyond the physical. I had been with Muktananda less than a year when he told me one day that I identified too much with my physical body. Like the smart-aleck that I can be, I asked him whose body he thought I should identify with. He ignored my quip and proceeded to expound on a theory I had heard before but largely dismissed: We have three levels of being, he said, physical, subtle, and causal. I was, he pointed out, aware of my physical body but not really aware of my subtle body and had no awareness at all of my causal body. "For the next week, spend half your morning meditation in the usual posture," he suggested, "then lie back into the supine, or corpse position, flat on the back."

A suggestion from Baba was a hefty command to me, so I did just that, starting the next morning. I generally spent two hours on this inner work anyway before the ashram moved into full swing at about five. The moment I lay back into the supine position, I went out of my body without any transition or warning. Going out of body is an unmistakable feeling—I had experienced variations of it before, during, for instance, one of Robert Monroe's mind-altering weekend programs in 1975.[2]

There in the ashram my body sensations were quite intact as I floated about six inches above my physical body. I clearly felt the heat of my body directly beneath me and heard it breathing. I didn't need to breathe, but

2. Gurumayi claimed we never actually go out of body but simply shift frequencies, an astute insight. Recall the observations of Mae Wan Ho in chapter 4.

with normal vision as well as hearing and feeling, I looked around the room in the dim predawn light and thought of rolling over and looking at my body, which is supposedly a risky thing to do. I couldn't manage it, though, and was largely helpless. I then thought I should have some great out-of-body venture, such as Robert Monroe had written about. (See Monroe's famous book, *Journeys Out of the Body*.) But nothing happened at all. It was a most uneventful event. After some half hour or so I felt myself settling back into my body, after which I got up and went about my day.

This happened for five straight mornings in the very same way, and all five times I thought I should will myself out into the wild blue yonder and have a bit of a blast. But no such luck. I was stuck in that stupid posture, floating on my back above the other part of me, five times. On the evening of my fifth time, in *darshan* (our daily meeting with the guru) the issue of not having some great out-of-body experience was hot in my roof-brain chatter, while Baba, through Gurumayi's interpretation, was giving his evening talk. Suddenly he broke off, looked straight at me, and said, through Gurumayi, "You can't go anywhere in your subtle body alone. It is too weak to move beyond its physical counterpart. To travel into other realms you must identify with your causal body while in the subtle body."

I realized then that the five straight repetitions of my out of body with no frills attached was simply Baba's gift. He had me experience my subtle body so thoroughly and unmistakably that I should not easily again forget or doubt. I could then have the adventure of discovering the causal body, which lay, according to Baba's cosmology, in the heart. As an aside, Rudolf Steiner developed quite a theory of subtle or ethereal states and bodies, and how to move in them, although to the rest of our culture doing so is an embarrassment. Between the church frowning on any kind of parapsychological experience and the scoffing of scientism, we don't give these notions much credibility and they remain weak and peripheral, as we ourselves do.

Baba overestimated my capacity and tenacity, however, because I lapsed back into my usual fearful state, concerned as ever over bodily harm and the specter of death. Eventually I found that when I was really in sync with my inner state, my outer fears would temporarily abate, rather as in unconflicted behavior. Muktananda's gift gave me concrete, personal knowledge of a level of my self-system lying beyond my usual physical awareness. This personal knowing—along with a series of events that occurred thirteen years before I met Baba—added to my conviction that some form of resurrection may well have taken place with Jesus.

After my first wife died, at age thirty-five, she made several dramatic returns to us, always in relation to and in the presence of her fifth and last child, who was about a year old when my wife died. Up to her last moment my wife was passionately intent on healing that infant, a victim of severe cerebral palsy. While the first events are far too involved to explain in full, suffice it to say that a few days after her death, she let her presence be known in two remarkable paranormal occurrences involving the infant. These were followed by a visit that was very powerful, as objects were moved around and the body of the infant was manipulated. All of these visits paved the way for what followed.

She then came to us in two manifestations that were altogether visible— she appeared in quite solid fashion. She first appeared at about ten in the evening and stood over the infant's crib for a surprisingly long time. In the second, my wife appeared, looked intently at the child, and then looked long and steadily at the child's grandmother, her mother, who planned to take the infant home with her and care for it while I took care of our four other children. In both these manifestations my wife appeared as she was when she was about twenty-two, dressed in her favorite pink suit she was married in and that she had carefully kept. Perhaps the form she assumed was a combination of her self-image, her mother's most lasting image, and my fondest memory of her. Each appearance lasted for what seemed a long while.

A third such manifestation occurred in New England a month or so later, where my wife's mother had taken the child. This time my wife came late in the night, standing between the infant's crib and the grandmother's bed. This immediate proximity drained all of my mother-in-law's body heat. She awakened nearly frozen and terror-stricken and later recounted to me how, after sensing her daughter's immediate presence again, she began to pray to her daughter fervently and silently to move away from so close a position. At this point my wife moved to the other side of the crib and her mother's body heat slowly returned. My mother-in-law called me on the phone as soon as she was able, at an ungodly late hour, and was seriously upset over the body heat business. She said I was the only person she dared tell because I was the only one who might believe her rather than send for the men in white coats. Actually, I was not too surprised, having read of such ghostly accounts in esoteric literature (and even in *Reader's Digest*).

At any rate, although these manifestations took place within a few weeks of her death and were not repeated after the last one at her mother's, I had experienced in quite a tactile and sensory fashion events that are not correct to talk about in our sane society and would surely be dismissed in scientific circles.

THE POWER OF PASSION AND COMPASSION

▼

I bring up this story of my wife's return to make a simple observation: If a present-day mother, driven by her extraordinary concern and passion for her damaged child, could break through the barrier of death and manifest in the way that she did, why should history's great model, driven by his passion for the whole of our species, not have done the same? The fact that such manifestations are products of our own visual system, as is the swinging sun at Medjagorge, is simply an example of Maturana and Varela's observation that the eyes see what the brain is doing even as the brain does according to what the eyes see. The creator-created dynamic is a function without boundaries, and at times we are graced with a breakthrough of our personal ones.

And so I say to all the modern theologians who apologize away the resurrection and the miracles and to the noise of literal fundamentalists who go to the opposite extreme beyond all common sense and so miss the point: It's not just that most of the reported miracles of Jesus can be found duplicated somewhere even today, in random, scattered fashion, but that they are actual examples of the human potential that we all possess. Deny them in Jesus and you surely deny them for us all. We quickly seal any cracks in our cosmic egg lest the unknown assail us, even when that unknown is an expression of our highest nature and what is known is killing us. To maintain our position of fear and victimization requires enormous expenditures of energy that could be employed otherwise.

Jesus and the intelligence of life did what they could to heal our fractured minds and hearts, and culture did what it had to do to squelch his magnificent gesture and make a religion of it, "a homeopathic remedy for his viral threat" to culture. Projected onto that mystical christ floating in the heavens, we can dismiss the reality of Jesus and his cross and the unconflicted nature of his faith. The whole operation can be moved into the ethereal realm of marshmallow make-believe and culture will remain supreme. There, I suppose, we can at least all believe and go down together, no doubt as a good, praying congregation begging mercy from that tyrannical "moral governor of the universe" and his "only begotten son," that equally victimized moral whip and judge of a victimized, fated species.

Or we can pick up that cross and reclaim our birthright; rescue Jesus from the Christians, bring him down from cloud nine, and find him reflected in our mirror; see him in each and every face on the street as Whitman did, find him even in the least of these our brethren behind bars. We too can risk ourselves; throw ourselves to the winds as he did; drop our fearful defenses, judgments, self-justification, shame, and guilt; and embrace that life of greater gifts that he displayed, performing, as he promised, even greater works than his, and so rise and go beyond.

TWELVE

▼

THE RESURRECTION
OF EVE

Error is created . . . It is Burnt up the Moment Men cease to behold it.

—WILLIAM BLAKE

Morris Berman, in his sobering and prophetic work, *The Twilight of American Culture,* recounts that with the collapse of the Roman Empire at the hands of the Visigoth barbarians, who overran and sacked what passed for civilization, monks in monasteries began collecting all artistic, philosophical, or religious treatises they could lay hands on in order to hide them for safekeeping. Thus it was that the literary treasures of Greek and Roman culture were preserved in the coming Dark Ages, to be discovered anew centuries later, sparking the so-called Enlightenment and Renaissance in Europe. Today, Berman argues, we need the same—a new monastic order that will harbor the elements of culture as we sink into that twilight he and many others foresee.[1]

Saving our cultural heritage may be a mixed bag at best; it would be the collections of a monastic order saved according to male notions of what is important. Perhaps, along with salvaging what artifacts we can from our existing culture, we would carry over the virus infecting it. We might, to better advantage, try saving our biological heritage, a well-worked-out package billions of years old that has behind it the intelligence of life and not just the intellect of a patriarchal fiasco. With our biological apparatus intact, we could create cultures at will, even benevolent ones, and let the past and its miserable bloodlust fade away.

1. Morris Berman, *The Decline of American Culture* (New York: W. W. Norton, 2000).

How, though, would we proceed were we to set out to save our biological heritage? First, consider Gurumayi's claim that the heart never "solves problems," but gives us a new situation, a new reality, if we will allow it. We have already covered one procedure for recapturing our lost intelligence of the heart, one based not on pie-in-the-sky New Age dreaming but on simple biological facts. Perhaps it is time, then, not for more intellectual engineering but for allowing a truly new reality to emerge from our true nature, a time to gather together all the strands or clues from neurocardiology; the new biology; the new physics; energy medicine; the findings of the Pre- and Perinatal Psychology and Health Association concerning conception, pregnancy and childbirth; and our new insights into the critical importance of the earliest developmental stages. We might weave all the disparate strands into a kind of Ariadne thread that could lead us out of our current maze of conflict and confusion. This is the time to carry forward our knowledge of the creator-created dynamic and what makes a complete human, our knowledge of the strength of love and forgiveness and the opening of heart. All this ancient yet new-to-us insight is pyramiding around us even as chaos mounts in equal measure. Surely this wild polarity makes ours the most interesting and exciting of times, perhaps the most potent and open-ended period yet in our human venture, particularly because, from our created end of the dynamic, it is up to us to ensure that out of the chaos of our collapsing culture the foundation of a new and positive reality emerges.

Nobelist Ilya Prigogine, the Belgian chemist, suggested that a system in balance and functioning well is difficult to change, but as a system falls into disorder, change becomes more and more feasible and finally inevitable. At that inevitable point the least bit of coherent order can bring to order the whole disorderly array. Which direction the change takes depends on the nature of the chaotic attractor that lifts the chaos into its new order—which is a variation on our old friend the model imperative. If that chaotic attractor is itself demonic, the old cycle simply repeats itself, which seems to have been historically the case for our species. But if the chaotic attractor were benevolent or "divine," the new order would have to be of that same nature. A positive outcome should have occurred two millennia ago, but perhaps the collapse and chaos at that time weren't thorough enough. As our history since then has shown, we had even further to sink before a real bottoming-out, and this just might be the time.

Consider the history of the Polish Solidarity, the name given a number of small labor groups rather isolated from each other during the Communist era. All the disparate groups shared the same coherent, clear concept of national economic and social reform. When the Communist system crumbled, Solidarity, having quietly gathered momentum, was seriously influential simply because of the catalytic effect of orderly thought in a time of disorderly chaos. Poland became the first former Iron Curtain country to achieve economic and social stability. As this example illustrates, the catalytic effect of a chaotic attractor is not a numbers game, but rather a matter of coherence amid incoherence.

In the gathering wave of new chaos, then, sensing the futility of trying to change monolithic structures of institutional thought and practice and recognizing that the ship of state may not be salvageable, we should work to build sound little lifeboats, Solidarity-style. A blueprint for those lifeboats has been sketched throughout this book, a plan based not on political or economic notions but on biological fact. Our concluding pages here will view this blueprint in a new light, making our close not just a recapitulation but an opening to something new and unknown.

Around the mid-twentieth century, some two hundred male medical students at Harvard University were interviewed to determine the extent or lack of parental nurturing they experienced in infancy and childhood. The subjects were grouped into positive and negative categories accordingly, those nurtured and those not. Forty years later the surviving men were given physical examinations. Of those who rated their parents supportive and nurturing, 25 percent had illnesses related to age. Of those rating their parents unsupportive, 89 percent had age-related illnesses.

Gary Schwartz and Linda Russek, of the University of Arizona, made a further test of a representative group of these men in this manner (my summary of that study is markedly abridged but essentially accurate): Each subject was wired for EEG (brain) and ECG (heart) frequencies and was seated three feet from the interviewer (Russek), who was herself wired in the same manner. Within a short time the averaged EEG (brain-wave) patterns of those subjects having positive childhoods synchronized or entrained with the averaged ECG (heart-frequency) patterns of the interviewer. (Recall from the beginning of chapter 3 the example of the two heart cells that markedly influenced each other on the microscope's slide, and the fact that the heart's em torus is quite strong within a three-foot radius.) The EEG patterns of the subjects with negative childhoods showed a much slower-forming and weaker correspondence to the interviewer, if any at all. (See figure 10 for a similar experiment at HeartMath regarding the "electricity of touch" between two people.) Recalling that the immune and emotional systems are of the same order, the implication here is that emotional deprivation in infancy and childhood predisposes an individual to a lifetime of essential loneliness or isolation, as well as to the attending susceptibility to disease. We learn to love by first being loved, and love seems the best armor against illness.

So the first order in lifeboat building is to recognize for whom the boat must be built—in this case, obviously, the child. A child's lifeboat, however, is made up of that child's creator and caregiving parent and/or parents. She who creates and brings the child into the world also models for him, educates him, and leads him forth into knowledge. She, then, must be one with the knowledge of that greatest and most priceless good news of who we are; one with our creator; and a principal part of the dynamic of creation, not a

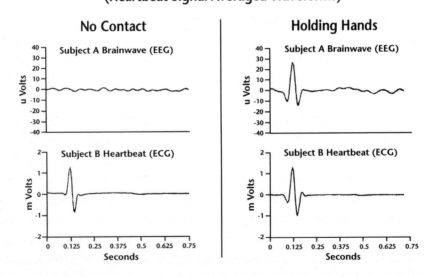

Electricity of Touch
(Heartbeat Signal Averaged Waveforms)

Figure 10. Heartbeat signal averaged waveforms showing a transference of the electrical energy generated by one subject's heart, which can be detected in the other subject's EEG (brain waves) when the two subjects hold hands. Courtesy of HeartMath Institute.

victim of it. Just as creator and created give rise to each other, so do parent and child. The intriguing thing about this dynamic is that the child awakens in the parent ancient vistas of knowing vital to the ongoing dynamic, in order that she might awaken the same in that child. Mirror to mirror again. So a truly new beginning must begin not with the child, but with the future mother, so that she may be awakened to the awareness of who she is, that she is in charge of her life—preferably before she conceives and critically before she gives birth. In this way, from conception to pregnancy, pregnancy to birthing, and birthing to life itself, the child and caregivers are able to mirror more and more in a spiraling gyre based on that love that is the foundation.

Surely fathers are indispensable for this sea change, but we must start with mothers and women at large. Males, it seems, have lost their moorings, leaving Plato's words more true today than ever: "Give me a new mother," he said, "and I'll give you a new world."

Laying the foundations of a new mind and new world has been her task from the beginning, and substitutes just haven't worked. Patriarchy has failed us. Recall the marvelous fairy tale of the noble king who falls into error, sinking down into his basest self to find himself locked in a beast's body from which he cannot extricate himself. Another tale relates how the handsome prince finds his erring ways have trapped him in the body of the lowly frog. From human to old mammalian to reptilian—to where now, when the only lower step is death itself? The resolution comes, our tales tell us, not from knights in shining armor and mighty exploits of strength and courage, or even the wisdom of sages and seers, but through the gentle gesture of the eternal She, whose nurturing kiss alone can save him from himself.

Glynda Lee Hoffman explored one of the most famous appearances of that eternal She in her book *The Secret Dowry of Eve*. In this brilliant study the author elaborates on a notion many have held concerning Eve's true role in that memorable myth of the Garden with its famous apple, infamous Serpent, and hapless Adam. With the ancient Kabbalah as guide, Hoffman spent some twenty years studying Genesis in the original Hebraic alphabet, which, she contends, throws new light on our literary and religious heritage. One observation from Hoffman is particularly pertinent to our needs here: Eve preceded Adam, of course (as any biologist would affirm), and she was the one granted by the Serpent not a curse, but instead the boon of conscious awakening. (Was this the first case of enlightenment?

Or possibly the last?) Awakened to her true nature, Eve, in turn, awakened Adam as best she could. To understand the full implications of this we need to recall that monolithic myth in stone, the Sphinx, and its towering triumph of the great serpent arising from the crown of the human skull.

After all, what is evolution and transcendence all about, and where more clearly depicted than in the Sphinx? The reptilian foundation on which human life is built is lifted into ever-greater orders of functioning. Again and again the higher incorporates the lower into its service, changing the nature of the lower into that of the higher, until that which was lowest is lifted to the highest, wherein we have risen and gone beyond all limitation and constraint: the resurrected human. So never sell that reptile short. He and Eve may have been in cahoots from the beginning. Who else can soothe the savage beast and lift him up but She? Understanding something of that serpent power curled over the insightful third eye and arousing the orbito-frontal loop to move evolution along, we can see why the Genesis God was jealous of this upstart couple who were now on a creative par with him, just, in fact, as the Serpent had promised Eve. These were big stakes.

So we can forget the many ways in which patriarchy inverted this magnificent tale of our beginnings—from that ridiculous Adam's rib nonsense down to the sentencing of Eve to great pain and turmoil in childbirth as punishment for her erring ways (a myth that took strong hold in our Judaic-Christian psyches and so proved to be the case, as can happen easily with negative imprints).

We do well to remember the primacy of Eve, she who birthed our species and gave us sight, and this concluding chapter is nothing less than a call to colors for all Eves to rally around a resurrection of your ancient forebear and work to bring a second enlightenment to this blind, sleep-walking world—particularly to us Adams in it, who didn't hang on to enlightenment too well the first time around, if it took at all.

I am not speaking of resurrecting the New Age fantasies of wild women who play goddess or priestess for new temples; or of female warriors who vision-quest or run with wolves. These are false substitutes that at best ape the behaviors of males. The real clarion call is for civilized women who will both birth and nurture a reborn species, altogether a far greater challenge than playing superman in a bra. The resurrection of Eve is the resurrection of that woman sacrificed to the altar of a patriarchal lineage now millennia old; that medieval woman lost in the labyrinths of fearful witch-hunting and crone-burning ecclesiastics; those midwives who, striving to nurture other women at their time of greatest need, have at times been jailed through the machinations of male lawyers directed by male lawmakers influenced by male doctors; those young mothers who have no choice but to submit to male-dominated hospital childbirth that disrupts the bond of love from the beginning. It is the resurrection of those financially poor mothers and their newborns who are left on the street with no help and are told by our government to get jobs and place their infants in daycare environments that might well compromise their children's well-being; or those mothers who are themselves abandoned by their men and left to struggle single-handedly to do that which is difficult in our time for an intact family to do—survive an economy that rides roughshod over the grist needed for its mill. Or, farther afield, it is the resurrection of those raped women singled out for their gender to receive the hate and rage bottled up in young soldiers incited or driven to the ultimate murder of war and its atrocities (read the conclusion of Gil Bailie's *Violence Unveiled,* if you have the strength for it); or those pubescent girls studied by Harvard's Carol Gilligan, who are found to be so confident, sure, and idealistic at eleven or twelve and are too-often defeated and depressed by fifteen.

The recruits for Eve's resurrection should rally from far and wide, as should the models needed for this resurrection, those that are equally far-flung and plentiful, if less visible. In the late 1980s, for instance, a three-year-old boy was brought by his mother to our ashram in India. At first glance I was sure the child was hydrocephalic—his head was of normal size and shape, but instead of an ordinary forehead, two large, bulbous semi-spheres extended well beyond the tip of his nose and flared out from the temples. There was an actual cleavage visible between the hemispheres, and a sharply dimpled indentation at the temples, marking the place where the enormous forehead blended into the rest of the brain case. He was a bright, calm, composed, and observant child with excellent motor skills and noticeably intelligent eyes that were both deeply penetrating and un-wavering. He steadily returned my gaze as if, it seemed to me, he was quietly sizing me up upon finding that I was staring at him.

His mother was a beautiful creature, one of those remarkable women who own themselves and exude inner security, confidence, and intelligence.

She was, it turns out, in her mid-forties, head of a worldwide spiritual organization, which wasn't hard to imagine, and the mother of three older children as well. This little boy, she informed me, had been deliberately conceived, cared for during pregnancy, and birthed in the ocean in the Bahamas, about as far from a hospital as you can get. Friends have told me that they recently met her son, now grown, through his mother's organization, and had found him an unusual and extraordinary young man.

Today I consider this child's birth significant, possibly the first or one of the first of a growing line of similar children who have been entering the world in recent years. I do believe nature has been responding to our need for a higher level of intelligence and spirit in giving us such children with such pronounced prefrontals. I spot these children occasionally, carried by their mothers in airports or at lectures in some variety of Snugli or baby sling ("baby-wearing"). I met such a child, a four-month-old boy, at a birthing conference in Thailand in 1998;[2] a three-year-old child at a yoga center in Connecticut who happened to be a remarkable African-American girl with an equally remarkable mother, just to squelch any supremacists notions that might emerge; and, just recently, a nursing infant in the audience at a lecture in Bellingham, Washington. On spotting such children I am drawn like a magnet. The gaze from their eyes is captivating, and

---

2. This four-month-old infant's mother was a Russian woman in her mid-forties with several other children. For years she had done prenatal and perinatal work with Igor Charkovsky, the famous water-birth doctor. Her boy had been born in the Black Sea and his birth had been attended by dolphins—a fairly common practice among the Rainbow Dolphin group in Australia, but surely oddball New Age monkey business to academic mind-sets. This four-month-old Russian infant, with prefrontals almost as pronounced as the child from New Mexico at the ashram, was the most precocious child I have ever witnessed. I could sympathize with the French physicians who, in the 1980s, published about "Charkovsky's babies" as they grew into adults, claiming they were so advanced and intelligent they were like people from another planet. While exceptional in every way, these children studied by the French did not exhibit these pronounced prefrontals until recently—as far as I know. Nature, then, may well be upping the ante even more in her bid to bail us out. Interestingly enough, these "Charkovsky babies" are, as grown people, essentially anonymous in society. They are apparently not hell-bent on making the front page, which may indicate a blessedly shallow enculturation.

once contact is made, you don't want to break it. Like Carlos Castaneda's coyote, they seem to tell us in that look "everything there is to know."[3]

Recall the research of 1998 showing how a mother who is emotionally mature, stable, loved, and feels secure gives birth to a child with an advanced forebrain, and Allen Schore's work showing that an infant protected and nurtured has a larger prefrontal growth after birth and maintains that growth during the toddler period if nurturing is unbroken.

The mothers of such children whom I have spoken with at any length have strikingly similar backgrounds. They are self-possessed women of strength and self-confidence and are deeply spiritual in a personal rather than formal sense. Many are in their late thirties or early forties.

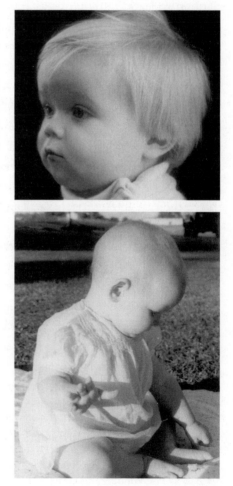

3. These children with pronounced prefrontals have nothing in common with those described in a New Age book concerning so-called indigo children.

Figure 11 shows two such children, a fifteen-month-old boy (top), and a six-month-old girl (bottom). Neither child has an exceptionally large head, but both do have foreheads extending beyond the tips of their noses, indicating extremely large prefrontal lobes. I have no figures on this but the number of such children seems to be increasing, not because of fewer bombs falling and our environment becoming safe and protective, but because more and more women are discovering their personal power and are able to create their own internal environment of peace and security regardless of the outer world. And this is all that counts. It is this very inner security that must be fostered at all costs.

Figure 11. It is easy to see that both the fifteen-month-old boy (top) and the six-month-old girl (bottom) have pronounced prefrontal lobes.

Their children were very much wanted; most were deliberately conceived; and all but one that I know of were born without medical intervention. All were bonded and breast-fed and sufficiently "worn" by the mother for the critical in-arms period.[4]

Because, as Blake claimed, anything capable of being believed is an image of truth, regardless of whether or not its authenticity can be nailed down for hardcore science, we might consider these few evolutionary modulations appearing among us as cues from nature, as inverse forms of the model imperative: Truly a little child shall lead us. Though these children are as yet rare, we have surely been given sufficient evidence to move vigorously ahead with Eve's resurrection. We might actually give nature greater opportunities to expand this as yet tiny vanguard.

The father's role and challenge, both before and after that spectacular bonding that plants the seed of new life, is to support the mother right down the line and provide her with a safe space that is free of fear so that the child's safe space, first within the mother and later with her, is never in question. To force a mother to fend for herself after giving birth, as is too often the case in America, exacts an awful social price all of us must pay.

Fathers are the bridge between nest and world at large and they are as important and subject to the same model imperative as are mothers in pregnancy, a child's infancy, and a child's first three years. (For those interested in a stunning model for fathers, I would heartily recommend David Albert's remarkable book, *And the Skylark Sings with Me,* which is an account of his experience helping to bring up and homeschool his precocious daughter.)

Children thrive under the protective umbrella of both mother and father, rare though this is becoming. But the American-style nuclear family was an accidental expedient perpetuated by corporate manipulation and

---

4. The one exception I know about is the boy in the photograph. He fits the template of the others in every respect except that as his birth unfolded one of those rare emergencies developed that the midwife could not handle (this occurs in .5 percent of all births) and medical intervention was necessary. He was C-sectioned and immediately given to his mother so that bonding and breast-feeding could proceed smoothly. Thus, when reserved for true emergencies, some tricks of modern obstetrics could be a blessing rather than a curse. In Holland, 95 percent of all children were born at home throughout the second half of the last century where they were delivered by a midwife team that traveled in a medical van equipped with all modern emergency devices. Holland had the lowest infant and maternal death rate in the world for decades.

state, religious, and political opportunists. Consider that in the 1890s roughly 94 percent of all Americans lived on farms where the extended family was the rule because it was economically expedient. One hundred years later, 96 percent of all Americans live in cities and towns, which is most expedient for corporate, political, or state-religious concerns, but is unviable and disruptive to the nuclear family. In these environments the nuclear family has been short-lived and rightly so. Michel Odent points out that the nuclear family by itself is an unnatural and nonviable relationship, but when the nuclear family is the nucleus of the extended family, and the extended family of society, the system works beautifully. If you strip away the extended family, however, as we have largely done, the nucleus implodes. Most of our legal nuclear couplings collapse and too many of those that hold up are of men and women "living lives of quiet desperation," as Thoreau would put it, enduring their lot for a raft of culturally imposed sanctions that rest on guilt, shame, and fear.

Consider Jesus' observation that "in that kingdom there is no giving and taking in marriage." This is an invitation neither to "free love" nor to celibacy, but a call for a form of relationship not contingent on legal contract. A cultural-religious marriage is a public vow, like an oath, to which Jesus was seriously opposed because oaths bind the spirit and close the open-ended nature of life. Further, cultural marriage is a legal contract in which each party unconsciously assumes that the other is now a possession, literally a property they own, from whom they have legally binding expectations and demands, which can lead to the equally disastrous notion that parents "own" their child. This notion of ownership is a major spiritual antagonist that the new Eve will be able to avoid. Who knows what kinds of fluid, living, vital relationships might unfold in a transcendent state, or what the needs may be of the child prepared for and nurtured by Eve?

To set up criteria that must be met in such a new society would be to replicate our cultural cul-de-sac, however, and is not my intent. There are, on the other hand, one or two preliminary moves that must be made to bring Eve to new life. First in resurrecting Eve is to ensure her rights over her own reproductive process, eliminating men, politics, and religion from what has been from time immemorial the female's prerogative. Second, she must reclaim her birth rights and body, which have, in both instances, been co-opted by men, the former by physicians and the latter by Madison Avenue and Hollywood.

It can be plainly stated: Get rid of male intervention in women's issues, particularly at birth—even in its new guise of female obstetricians—and eliminate completely the delivery room arena. Birthing is an ancient mammalian intelligence, not a problem to be solved through masculine intellect. All delivering mammals seek out the most private, quiet, safe, and generally dark place available for giving birth, which is, after all, the most private and wondrous creative act. But it is also one in which the mother is most vulnerable. In all mammals, including humans, at the first indication of some interference during birthing, the slightest foreign noise indicating possible intrusion, millions of years of genetic encoding will shut down the birth process, and it will be put on hold until the coast is clear (to a point, of course).

More and more women are discovering the joy and liberation of delivering at home, and even more the joy of solo birthing, with a midwife well in the wings but ready if need be. Unmolested and secure mothers can give birth in as little as twenty minutes, and are filled with both strength to nurture immediately and an equal joy in so doing. Eve has millions of years of genetic encoding built in to guide and direct her in that great venture.

The answer to our present dilemma, then, lies in prevention of current error, not in therapy, and the challenge for Eve's new life is to do things from the beginning according to nature. Natural function, allowed to unfold during its window of opportunity, is far cheaper than later compensation. Bringing a child into the world through a conception, pregnancy, and birth in keeping with nature's agenda gives our great mother, nature herself, the opportunity to lift us beyond our restraints and obstacles. And even now she responds at every opportunity by providing the neural capacities needed. How wide the gates of transcendence might open were great nature given unlimited opportunity—and Eve can make it happen!

FLING WIDE THE GATES

▼

Eve's revolution will lift sexuality back into its realm of greatness, for sexuality is a gateway to transcendence. Closed, it diverts us to violence and hell. Religion's branding of sex as a primal sin was a primal sin itself and a direct attack against Eve. All state-religions are guilty of this, East and West. Under such baleful influence even our so-called liberated sex of the sixties became a bag of worms leading us into more turmoil and psychological anguish. We cannot know what new codes would manifest in a benevolent world, nor do we need to know to make the first great steps toward reaching it.

Again, the ball is in Eve's court and her revolution is the radical break with culture that will save us. A true break with culture involves nothing other than picking up the cross. What is involved in that way of Jesus will unfold by default in this revolution: We will drop our deadly defenses, judgments, self-justifications; we will leave behind self-pity, retribution, demands for justice, and fearful reactions that lead to law and war; we will cease ruling out love. Noble vows, statements of belief, and creeds have failed, as has the patriarchy that invented them. Picking up the cross shifts us out of hindbrain survival instincts and opens us to the higher frequencies of love, forgiveness, and trust. Intellect will open to heart and move for the well-being of life.

We will all find, upon picking up that cross, that its burden is light, for the crushing load of enculturation is gone. And in the freedom of the unconflicted behavior that remains, our journey into God will open wide.

EPILOGUE

Robert Wolff's book *Original Wisdom* is a memorable account of his younger years spent among the Malay people in the early and middle parts of the twentieth century, a people who are now essentially extinct culturally (though not physically), and his life among the elusive, indeed near mythical Senoi, who were in Wolff's time a scant handful of aboriginal people living without restraint or law in the remaining Malay jungle. Unless they chose to be seen, the "primitive" Senoi were, in effect, invisible and unheard in the clamor of encroaching culture, a presence that the Senoi knew would eventually spell their end. The Malaysian government of the time was clearing the jungle for rubber plantations and the Senoi knew they could not live without the trees that encompassed their world and communed with them on many levels.

These people made up a society of benevolence and what we would call unconditional love, though I doubt they had a word for or could grasp the concept of love any more than a fish could grasp the concept of water. What else is there? a fish would ask. We most often coin words for that which seems other to us; it seems that in lacking something we label it, thereby creating a semantic substitute. The Senoi lived with unquestioned acceptance of each other, without judgment or censure, in a natural and spontaneous manner that was simply the only response they knew.

A hallmark of the Senoi was their unbroken, silent communion with their environment and each other, an integrated, self-contained way of relating that needed no reference to anything outside itself, and that did not, therefore, lend itself to analysis or description by an outsider. Walter Stace wrote about the "extraverted mystical experience," a fusion of heart and mind that occurs in a waking state and encompasses self and nature as an undifferentiated unit. This might describe how the Senoi lived—with a level of awareness beyond our comprehension; with a quality of being, a quiet steady joy, unknown to us; and with capacities of mind we can't grasp. Senoi life was constant divine play, unavailable and invisible to those caught in grim necessity.

Most of us, unable to play, are equally unable to recognize divine play when it takes place. Perhaps only Eckhart has described something similar to such "wandering joy" as the Senoi displayed—though coming from a radically different worldview and heritage, he and the Senoi may have opened to the same wavelength, along with, perhaps, Blake, who wrote:

> *How do you know but every bird*
> *that wings the airy way,*
> *is an immense world of delight*
> *closed to your senses five?*

Such divine play and wandering joy opens only through freedom from judgment and its resulting guilt, and/or restraint with its accompanying emotional and physical blocks. The Senoi refrained from judging self or others not from some noble virtue but because their minds, not having been formed in the same manner as ours, simply didn't function that way—never having been judged or restrained, they had no concept of either and no neural paths for relating in these ways. We, on the other hand, having been restrained and judged since birth, automatically judge others, restrain them if possible, and teach our children to do the same.

Not judging the actions of ourselves and others and trying to modify behaviors accordingly may seem negligent to us, but to the Senoi a person's actions were simply the given of a situation, like the direction of the wind or the slant of the sunlight. This mind-set, embodied in Jean Piaget's description of early childhood as "the unquestioned acceptance of the given," Eckhart's "living without a why," J. Krishnamurti's "choiceless awareness," Jesus' "kingdom" of relationship, and Matthew Fox's original blessing, is a state of mind that can open us to the higher functions of our forebrain while freeing us from enslavement to the hindbrain—a shift that wholly changes perception.

The Senoi lived the Sermon on the Mount, a heretical notion since no missionary had ever found them to preach it. My claim is that the Senoi prove conclusively that the kingdom to which Jesus refers in his sermon is our genetic "home," our true and natural state, and that it existed among the Senoi because it was not usurped by culture. This is not to say the Senoi were religious, but rather that they had no need for religion, which, along with culture, is nothing more than a means of social control.

We have difficulty accepting, even as hypothesis, that had we not been enculturated from the beginning of our life, law and restraint would not be needed. Yet this is the simple fact that the Senoi display. They reveal the lie in our cultural belief that without restraint humans are beastly. They prove quite conclusively that actions against the well-being of another are not in our genetic repertoire, but are instead conditioned responses. Ironically, such conditioning or learned effects, based on disrupting our natural development, leave us in the thrall of our survival instincts. The further irony is that such instincts in and of themselves—as they exist in an animal, for instance—are not destructive until they are linked to a crippled human intellect.

We have long recognized a clear connection among language, conceptual and perceptual systems, and culture. Jean Liedloff takes on the exploration of this relationship. She lived among the Yequana, an aboriginal group much like the Senoi, on the upper Cuara River basin in Venezuela and wrote about them in her book *The Continuum Concept*. She reported that these people had no word for disobedience, and she found it impossible to explain such a condition or phenomenon to them because no Yequana child had ever disobeyed. Their understanding seemed to include only that a child acted like a child—his actions were accepted unconditionally, for they had never seen a child not act like a child. To the Yequana, whatever a child did was what children do and each child did, though quite individually and uniquely, as adults did. The individual was simultaneously the generic.

No child had ever acted in a way injurious to or unpleasant toward another because no adult acted in such a way. Yequana children, like ours, lived in their parents' "unconscious," as Jung would have called it, and this unconscious was not built on the labyrinth of restriction, fear, and shame that we as children had to work through to make us "behave," and that we then inflict on our own children.

Yet Yequana children obeyed their parents and elders, in our sense of the word—immediately, completely, and as naturally as breathing, accepting unconditionally their parents' requests just as their parents unconditionally accepted their children's actions. An enculturated mind doesn't grasp what Jesus meant by saying "judge not that you be not judged," that as we judge so are we judged, that it all amounts to another way of reaping what you sow.

Concerning the Yequana treatment of children, Liedloff writes:

The notion of ownership of other persons is absent among the Yequana. The idea that this is "my child" or "your child" does not exist. Deciding what another person should do, regardless of his age, is outside the Yequana vocabulary of behaviors. There is great interest in what everyone does, but no impulse to influence—let alone coerce—anyone. . . . But where his help is required, he [the child] is expected to comply instantly. Commands like "Bring some water!" "Chop some wood!" "Hand me that!" or "Give the baby a banana!" are given with the same assumption of innate sociality, in the firm knowledge that a child wants to be of service and to join in the work of his people. No one watches to see whether the child obeys—there is no doubt of his will to cooperate. As the social animal he is, he does as he is expected and to the very best of his ability.[1]

The Yequana children live out the parents' assumptions concerning them precisely as our "terrible twos" and later "terrible teens" live out ours, often to our dismay. Like our children, Yequana children become as they behold. "Do as I say, not as I do," we command, in contrast to the Yequana, who do first and never need to command. Their modeling actions are imprinted automatically.

That our children become who we are, more or less, rather than what we tell them to be, is a fact that can enrage us. But the model imperative is not a cultural invention subject to culture's modifications—it simply functions, like gravity. We have ignored for half a century or more the studies that show some 95 percent of all a child's learning or "structures of knowledge" form automatically in direct response to interactions with the environment, while only about 5 percent form as a result of our verbal teaching or intellectual instruction. The Senoi and Yequana are living proof of this.

We condescendingly observe that such people as the Senoi and Yequana do not "progress"—in fact, we say, they do nothing at all worthwhile that we can see. They create nothing, build or invent nothing new, and even neglect to develop their natural resources. Neither do they wantonly kill, rob, exploit, plunder, ruin their earth, or drive their children to suicide. Here's what the Senoi managed to accomplish: They lived a richly creative and utterly spontaneous life expressed in harmony within their own world. We cannot know what they have accomplished. We would have to be in

---

1. Liedloff, *The Continuum Concept*, 90–91.

their world with their eyes to experience the joyful novelty of shared adventure and wonder that they created anew each day. The Senoi demonstrated first-class forebrain stuff far beyond primitive R-brain instincts.

The Yequana lived a far more visibly organized and ritualized life than the Senoi, turning the most mundane matters of daily existence, such as drawing water, into dramatic and artistic pageantry enacted by each with individual style, flair, and grace. By turning what we consider work into play, they knew enjoyment from everything.

The Senoi and Yequana may well have been the foundation of what nature intended for her next evolutionary move, living in a way modeled by our great beings. How this marvelous plan became so diluted, lost, or distorted over most of our world is anyone's guess. Nor is there an answer for why, with increased intensity for the last century or so, we have been intent on destroying all traces of this aboriginal way of life. Jesus, our greatest being, said we must become again as an innocent, trusting child, not to remain in some permanent state of dependence, but as the foundation for greater things to come, a maturity beyond our present grasp. These greater things are possible, however, only through that child's state of unquestioned acceptance through which newness pours forth.

Consider again Paul MacLean's family triad of needs. The trust or faith in life that the Yequana and Senoi display is the hallmark of play, a play that emerges from true communication and nurturing. Because individuals in both groups had appropriate audiovisual communication since birth and lifelong nurturing, they served as living models of what play and becoming a child and playing can mean. Our world today looks on play as a waste. Frugal hard work is the ideal. That we are born to enjoy life is anathema to church, government, and industry. We must win the luxury of enjoyment through hard work—or purchase it on credit. In truth, in dying, we who have worked so hard and never played will be just as dead as a dead Senoi who lived his life in joyful play.

Because they lived in a reality that is not available to or registered by our "senses five," we have no more idea of who, what, or even where the Senoi were than we have of Blake's bird on the wing. Nor was such awareness available to Wolff until his egg was cracked through sufficient contact with a specific Senoi model. Then, in striking similarity to Castaneda and don Juan, Wolff finally dropped his defenses, self-importance, and general fear-based mind-set and opened himself to the Senoi way. The unique quality of Wolff's Senoi model was that speech played no part in the learning and metanoia that he shared. The state of the Senoi was the Tao itself, which doesn't lend itself to description.

Few of us have such an opportunity or the time for such an experience and even if we did, we would likely look in vain for a model like Wolff's. They are around, I am told, for those with eyes to see, but we are too busy surviving our culture, paying rent or mortgage and staggering insurance bills to protect ourselves against lawsuits from neighbors or damage by other people.

A life without judgment or restraint, and so without violence and law, can unfold only from a life without fear as modeled by Jesus or Peace Pilgrim, for instance. The extent or depth of our fear is largely a conditioned response, not a natural one, as George Jaidar (author of the intriguing book *The Soul, an Owner's Manual*) showed me years ago. An irony of history is that a child conceived, birthed, and brought up without restraint of any conceivable kind would never need restraining, not as a child, adolescent, or adult. This fact is Wolff's and Liedloff's gift to us. They clearly show how our natural state is one of unbroken relationship with our creator, in which everything works together for good as proposed, and that the natural instinct of the child is to maintain that state of relationship at all costs.

For such maintenance the word or concept *God* is not needed. God is not a semantic proposition or imaginative invention of our verbal brain to be believed in, as religion tells us, but is the force within us expressed as our very love of life and passionate will to live. As children we resist with all our will the loss of that original force bubbling up from within us, and this is the will that culture, particularly Moslem and fundamental Christian cultures, must—and does—break at all costs.

Restraint creates the necessity for restraint, and as it is increased, more is needed. Paul was right: Without law there is no sense of guilt or shame. But were there no guilt and shame, law and restraint would never have been conceived because they would not have been needed. A human nurtured instead of shamed and loved instead of driven by fear develops a different brain and therefore a different mind—he will not act against the well-being of another, nor against his larger body, the living earth. As a child we know we are an integral part of the continuum of all things, as Liedloff explains and Jesus demonstrated. We can and must rediscover that knowing.

RESOURCES

HeartMath Institute
14700 West Park Avenue
Boulder Creek, CA 95006
Phone: (831) 338-8500
Fax: (831) 338-8504
e-mail: info@heartmath.org
Web site: www.heartmath.org

SYDA Foundation
371 Brickman Road, P.O. Box 600
South Fallsburg, NY 12779-0600
Phone: (845) 434-2000
Web site: www.siddhayoga.org

Linda G. Russek, Ph.D., and Gary E. Schwartz, Ph.D.
Department of Psychology
University of Arizona
P.O. Box 210068
Tucson, AZ 85721
Phone: (520) 621-5497
Fax: (520) 621-9306
e-mail: gschwart@ccit.arizona.edu

BIBLIOGRAPHY

BIRTH AND BONDING

Ainsworth, Mary D. "Deprivation of Maternal Care: A Reassessment of Its Effects." *Public Health Papers* no. 14: 97–165.

———. *Infancy in Uganda.* Baltimore: John Hopkins University Press, 1967.

———. "Patterns of Attachment Behavior Shown by the Infant in Interaction with His Mother." *Merrill-Palmer Quarterly* 10 (1964): 51–58.

Arms, Suzanne. *Immaculate Deception: A New Look at Women and Childbirth in America.* Boston: Houghton Mifflin, 1975.

Bernard, J., and L. Sontag. "Fetal Reactions to Sound." *Journal of Genetic Psychology* 70 (1947): 209–10.

Bower, T. G. R. "The Visual World of the Infant," *Scientific American,* December 1966.

Bowlby, John. "Separation Anxiety." *International Journal of Psychoanalysis* 41 (1960): 89–113.

Chamberlain, David. *Babies Remember Birth.* Los Angeles: Jeremy P. Tarcher, 1988.

Cheek, David B. "Prenatal and Perinatal Imprints: Apparent Prenatal Consciousness as Revealed by Hypnosis." *Pre- and Perinatal Psychology Journal* 1 (Winter 1988): 97–110.

Condon, W., and Louis Sander. "Neonate Movement Is Synchronized with Adult Speech: Interactional Participation and Language Acquisition." *Science,* January 1974, 99–101.

DeChateau, Peter, and Brit Wilberg. "Long-Term Effect on Mother-Infant Behavior of Extra Contact During the First Hour of Post-Partum." *Acta-Paediatrix* 66 (1977): 137–43.

Fantz, Robert L. "The Origin of Form Perception." *Scientific American,* May 1961.

———. "Pattern Vision in Young Infants." *Psychological Review* (1958): 43–47.

Klaus, Marshall. "Maternal Attachment: Importance of the First Post-Partum Days." *New England Journal of Medicine* 9 (1972): 286.

Leboyer, Frederick. *Birth Without Violence*. Rochester, Vt.: Healing Arts Press, 2002.

———. *Loving Hands: The Traditional Art of Baby Massage*. New York: Newmarket Press, 1976.

Verny, Thomas, with John Kelley. *The Secret Life of the Unborn Child*. New York: Simon and Schuster, 1981.

THE BRAIN

Austin, James H. *Zen and the Brain*. Cambridge, Mass.: MIT Press, 1998.

Brenner, D., S. J. Williamson, and L. Kaufman. "Magnetic Fields in the Brain." *Science*, October 31, 1977, 480–81.

Edelman, Gerald. *Neural Darwinism: The Theory of Neuronal Group Selection*. New York: Basic Books, 1987.

Goldberg, Elkhonon. *The Executive Brain: Frontal Lobes and Civilized Mind*. New York: Oxford University Press, 2001.

Krasnegor, Norman A., ed. *Development of the Prefrontal Cortex: Evolution, Neurobiology, and Behavior*. Baltimore: Paul H. Brookes, 1997.

MacLean, Paul. "The Brain and Subjective Experience: Question of Multilevel Role of Resonance." *Journal of Mind and Behavior* 18, no. 2 and 3 (Spring/Summer): 247–68.

———. *A Triune Concept of the Brain and Behavior*. Clarence M. Hincks Memorial Lecture Series. Edited by D. Campbell and T. J. Boag. Toronto: University of Toronto Press, 1973.

———. *The Triune Brain in Evolution*. New York: Plenum Press, 1990.

———. "Women: A More Balanced Brain?" *Zygon Journal of Religion and Science* 31, no. 3 (September 1996).

Mark, George S., et al. "Mood Improvement Following Daily Left Prefrontal Repetitive Transcranial Magnetic Stimulation in Patients with Depression: A Placebo-Controlled Crossover Trial." *American Journal of Psychiatry* 154 (December 12, 1997).

DEVELOPMENT

Anderson, Stephen W., et al. "Impairment of Social and Moral Behavior Related to Early Damage in Human Prefrontal Cortex." *Nature Neuroscience* 2, no. 11 (November 1999).

Caplan, Mariana. *Untouched: The Need for Genuine Affection in an Impersonal World*. Prescott, Ariz.: Hohm Press, 1998.

Epstein, Herman T. "Phrenoblysis: Special Brain and Mind Growth Periods: I. Human Brain and Skull Development. II. Human Mental Development." *Developmental Psychology.* New York: John Wiley and Sons, 1974.

————. "Brain Growth Spurts" and "On the Beam." *New Horizons for Learning* 1, no. 2 (April 1981).

Healy, Jane. *Endangered Minds: Why Our Children Don't Think.* New York: Touchstone, 1991.

————. *Failure to Connect: How Computers Affect Our Children's Mind—for Better and Worse.* New York: Touchstone, 1999.

Jones, Blurton N. *Ethological Studies of Child Behavior.* New York: Cambridge University Press, 1972.

Luria, Alexander R. *The Role of Speech in Normal and Abnormal Behavior.* New York: Liveright, 1961.

Mikulak, Marcia. *The Children of a Bombara Village.* Santa Fe, N. Mex.: Santa Fe Research, 1991.

Piaget, Jean. *The Child's Conception of the World.* New York: Humanities Press, 1951.

————. *Intelligence and Affectivity: Their Relationship During Child Development.* Palo Alto, Calif.: Annual Reviews, Inc., 1981.

Prescott, James W. "Body Pleasure and the Origins of Violence." *The Futurist,* April 1975.

————. "The Origins of Human Love and Violence." Institute of Humanistic Science Monograph. 7th International Congress, Association for Pre- and Perinatal Psychology and Health, 1997.

Rody, Sylvia, and S. Axelrod. *Anxiety and Ego Formation in Infancy.* New York: International Universities Press, 1970.

Schore, Allan N. *Affect Regulation and the Origin of the Self: The Neurobiology of Emotional Development.* Hillsdale, N.J.: Lawrence Erlbaum Associates, 1994.

————. "The Experience-Dependent Maturation of a Regulatory System in the Orbital Prefrontal Cortex and the Origin of Developmental Psychopathology." *Development and Psychopathology* 8: 55–87.

Staley, Betty. *Between Form and Freedom.* Wallbridge, Stroud, England: Hawthorne Press, 1988.

Sweet, Win, and Bill Sweet. *Living Joyfully with Children.* Lakewood, Colo.: Acropolis Books, 1997.

THE HEART

Armour, J. A., and J. Ardell, eds. *Neurocardiology.* New York: Oxford University Press, 1994.

Cantin, Marc, and Jacques Genet. "The Heart as an Endocrine Gland." *Scientific American,* February 1986.

Childre, D. L. *FreezeFrame: Fast Action Stress Relief.* Boulder Creek: Calif.: Planetary Publications, 1994.

Childre, Lew, and Howard Martin. *The HeartMath Solution.* San Francisco: HarperSanFrancisco, 1999.

Childre, Lew, Rollin McCraty, and Deborah Rozman, eds. "Increasing Coherence in the Human System: A New Biobehavioral Technology for Increasing Health and Personal Effectiveness." Boulder Creek, Calif.: HeartMath Research Center.

Lacey, John, and Beatrice Lacey. "Two-Way Communication between the Heart and the Brain: Significance of Time within the Cardiac Cycle." *American Psychologist* (February 1978): 99–113.

Marinelli, Ralph, et al. "The Heart Is Not a Pump: A Refutation of the Pressure Propulsion Premise of Heart Function." *Frontier Perspectives* 5, no. 1 (Fall/Winter, 1995).

McArthur, David, and Bruce McArthur. *The Intelligent Heart.* Virginia Beach, Va.: A.R.E. Press, 1997.

McCraty, Rollin, ed. *Research Overview: Exploring the Role of the Heart in Human Performance.* Boulder Creek, Calif.: HeartMath Research Center, 1997.

Russek, Linda G., and Gary E. Schwartz. "Energy Cardiology: A Dynamical Energy Systems Approach for Integrating Conventional and Alternative Medicine. *Advances: The Journal of Mind-Body Health* 12, no. 4 (Fall 1996).

———. "Interpersonal Heart-Brain Registration and the Perception of Parental Love: A 42-Year Follow-up of the Harvard Mastery of Stress Study." *Subtle Energies* 5, no. 3 (1994): 195.

Song, L. Z., Gary E. Schwartz, and Linda G. Russek. "Heart-Focused Attention and Heart-Brain Synchronization: Energetic and Physiological Mechanisms." *Alternative Therapies in Health and Medicine* 4, no. 5 (September 1998).

RELIGION

The New English Bible, Oxford: Oxford University Press, 1961.

The New Testament, Revised Standard Version, New York: American Bible Society, 1952.

Chittick, William C. *The Sufi Path of Knowledge.* Albany, N.Y.: SUNY Press, 1989.

de Caussade, Jean-Pierre. *The Sacrament of the Present Moment.* San Francisco: Harper and Row, 1982.

Feuerbach, Ludwig. "The Essence of Christianity." In *The Young Hegelians: An Anthology*. Edited by Lawrence S. Stepelvich. Atlantic Highlands, N.J.: Humanities Press, 1983.

Girard, René. "The Bible's Distinctiveness and the Gospel." *The Girard Reader*.

Muller-Ortega, Paul Eduardo. "The Triadic Heart of Siva." *Kaula Tantricism of Abhinavagupta in the Non-dual Shaivism of Kashmir*. Albany, N.Y.: SUNY Press, 1989.

Pagels, Elaine. *The Gnostic Gospels*. New York: Random House, 1979.

———. *The Origins of Satan*. New York: Random House, 1981.

SCIENCE

Bateson, Gregory. *Mind and Nature: A Necessary Unity*. New York: E. P. Dutton, 1979.

Bohm, David. *Wholeness and the Implicate Order*. London: Routledge and Kegan-Paul, 1979.

Damasio, Antonio R. *Descartes' Error: Emotion, Reason and the Human Brain*. New York: G. Putnam and Sons, 1994.

Denton, Michael. *Nature's Destiny*. New York: Simon and Schuster, 1998.

Gardener, Howard. *The Mind's New Science*. New York: Basic Books, 1985.

———. *Frames of Mind*. New York: Basic Books, 1983.

Hannaford, Carla. *Smart Moves: Why Learning Is Not All in the Brain*. Baltimore: Great Ocean Press, 1995.

———. "Belonging from the Heart: The Science Behind Playing, Learning, and Living Coherently" (forthcoming).

Jahn, Robert G., and Brenda J. Dunne. *Margins of Reality: The Role of Consciousness in the Physical World*. San Diego: Harvest Books, Harcourt Brace Jovanovich, 1987.

Kafatos, Menas, and Robert Nadeau. *The Conscious Universe: Part and Whole in Modern Physical Theory*. New York: Springer-Verlag, 1990.

Kosambi, D. D. "Living Prehistory in India." *Scientific American* 216, no. 2, February 1967.

Maturana, Humberto R., and Francesco J. Varela. *The Tree of Knowledge: The Biological Roots of Human Understanding*. Cambridge: New Science Library, 1987.

Odent, Michel. *The Scientification of Love*. London, New York: Free Association Books, 1999.

Pert, Candace. *Molecules of Emotion*. New York: Simon and Schuster, 1999.

Raloff, Janet. "EMF's Biological Influences: Electromagnetic Fields Exert Effects on and through Hormones." *Science News* 153, January 1998.

———. "Electromagnetic Fields May Trigger Enzymes." *Science News* 153, February 1998.

———. "Magnetic Fields Can Diminish Drug Action." *Science News* 152, November 1997.

Russek, Linda G., and Gary E. Schwartz. "Interpersonal Registration of Actual and Intended Eye Gaze: Relationship of Openness to Spiritual Beliefs and Experiences." *Journal of Scientific Exploration,* June 2, 1998.

———. "The Origin of Holism and Memory in Nature: The Systemic Memory Hypothesis." *Frontier Perspectives,* 1998.

———. "Do All Dynamical Systems Have Memory? Implications of the Systemic Memory Hypothesis for Science and Society." In *Brain and Values: Behavioral Neurodynamics,* edited by K. H. Pribram and J. S. King. Hillsdale, N. J.: Lawrence Erlbaum Associates, 1997.

Selye, Hans. *The Stress of Life.* New York: McGraw Hill, 1965.

Sheldrake, Rupert. *A New Science of Life: Morphic Resonance.* Rochester, Vt.: Park Street Press, 1995.

Spinney, Laura. "The Unselfish Gene: Culture Shapes Our Evolutionary Destiny as Surely as DNA." *New Scientist* 26, October 1997.

Storfer, Miles. *Intelligence and Giftedness: The Contributions of Heredity and Early Environment.* San Francisco: Jossey-Bass, 1989.

Tiller, William W., Walter E. Dibbie Jr., and Michael J. Kohane. "Exploring Robust Interactions between Human Intention and Inanimate/Animate Systems." Paper presented at the conference Toward a Science of Consciousness: Fundamental Approaches, United Nations University, Tokyo, Japan, May 1999.

Weaver, Warren. *Science and Imagination: Selected Papers.* New York: Basic Books, 1967.

SPIRITUALITY

Carse, James P. *Breakfast at the Victory: The Mysticism of Ordinary Experience.* San Francisco: HarperSanFrancisco, 1994.

———. *Finite and Infinite Games.* New York: Ballentine Books, 1987.

———. *The Gospel of the Beloved Disciple.* San Francisco: HarperSanFrancisco, 1997.

———. *The Silence of God: Meditations on Prayer.* San Francisco: HarperSanFrancisco, 1995.

Fox, Matthew, ed. *Passion for Creation: The Earth-Honoring Spirituality of Meister Eckhart.* Rochester, Vt.: Inner Traditions, 2000.

Fox, Matthew. *Meditations with Meister Eckhart.* Santa Fe, N.M.: Bear and Co., 1983.

Franck, Frederick. *To Be Human Against All Odds.* Berkeley: Asian Humanities Press, 1991.

Harrison, Steven. *Doing Nothing: Coming to the End of the Spiritual Search.* New York: Tarcher-Putnam, 1977.

Hartmann, Thom. *The Prophet's Way: A Guide to Living in the Now.* Rochester, Vt.: Park Street Press, 2004.

Herrigel, Eugen. *Zen in the Art of Archery.* New York: Pantheon Books, 1953.

Muktananda, Swami. *The Play of Consciousness.* South Fallsburg, N.Y.: Siddha Foundation, 1980.

Myss, Caroline. *Anatomy of the Spirit.* New York: Harmony Books, 1996.

Roberts, Bernadette. *The Experience of No Self.* Albany, N.Y.: SUNY Press, 1993.

———. *The Path to No Self.* Albany, N.Y.: SUNY Press, 1991.

———. *What Is Self?* Austin, Tex.: Goen, 1996.

Segal, Suzanne. *Collision with the Infinite.* San Diego: Blue Dove Press, 1996.

Sells, Michael A. *Mystical Languages of Unsaying.* Chicago: University of Chicago Press, 1994.

Steiner, Rudolf. *Knowledge of the Higher Worlds, and Their Attainment.* London: Steiner Press, 1969.

Tweedie, Irina. *The Chasm of Fire.* Dorset, England: Element Books, 1979.

GENERAL

Alouf, Michel M. *History of Baalbek,* 15th ed. Beirut: American Press, 1937.

Berman, Morris. *The Twilight of American Culture.* New York: W. W. Norton, 2000.

Colavito, Maria. *The Heresy of Oedipus and the Mind-Mind Split.* London: Edwin Mellen Press, 1995.

Csikszentmihalyi, Mihaly. "Play and Intrinsic Rewards." *Journal of Humanistic Psychology* 13, no. 3 (Summer 1975).

deBeauport, Elaine. *The Three Faces of Mind.* Wheaton, Ill.: Quest Books, 1996.

Eliade, Mircea. *Cosmos and History: The Myth of the Eternal Return.* New York: Harper Torchbooks, 1959.

———. *Yoga: Immortality and Freedom.* Bollingen Series LVI. New York: Pantheon, 1958.

Feinberg, Leonard. "Fire-walking in Ceylon." *Atlantic Monthly,* May 1959.

Feuerstein, Georg. *Structures of Consciousness.* Lower Lake, Calif.: Integral, 1987.

Gebser, Jean. *The Ever-Present Origin.* Athens, Ohio: Ohio University Press, 1987.

Grosvenor, Donna and Gilbert Grosvenor. "Ceylon." *National Geographic Magazine* 129, April 1966.

Jensen, Adolf E. *Myth and Cult Among Primitive Peoples.* Chicago: University of Chicago Press, 1963.

Jung, Carl G. *The Archetypes and the Collective Unconscious.* Bollingen Series IX, no. 1. New York: Pantheon.

Katz, Richard. *Boiling Energy: Community Healing among the Kalahari Kung.* Cambridge: Harvard University Press, 1982.

Krechmal, Arnold. "Fire-Walkers of Greece." *Travel* 108, October 1957, 46–47.

Langer, Suzanne. *Philosophy in a New Key.* Cambridge: Harvard University Press, 1942.

Laski, Margharita. *Ecstasy: A Study of Some Secular and Religious Experiences.* Bloomington, Ind.: Indiana University Press, 1962.

Liedloff, Jean. *The Continuum Concept.* Reading, Mass.: Addison Wesley Press, 1977.

Lowndes, Florin. *Enlivening the Chakras of the Heart: The Fundamental Spiritual Exercises of Rudolf Steiner.* London: Sophia Books, Rudolf Steiner Press, 1988.

Mander, Jerry. *Four Arguments for the Elimination of Television.* New York: William Morrow, 1977.

McDermott, Robert A., ed. *The Essential Steiner: The Basic Writings.* San Francisco: Harper and Row, 1984.

Monroe, Robert. *Journeys Out of the Body.* New York: Doubleday, 1973.

Montagu, Ashley. *The Concept of the Primitive.* New York: Free Press, 1968.

———. *The Natural Superiority of Women.* New York: Macmillan, 1968.

———. *Touching: The Human Significance of the Skin.* New York: Harper Collins, 1978.

Murphy, Michael. *The Future of the Body.* New York: Tarcher-Putnam, 1992.

Pearce, Joseph Chilton. *Spiritual Initiation and the Breakthrough of Consciousness: The Bond of Power.* Rochester, Vt.: Park Street Press, 2003.

———. *The Crack in the Cosmic Egg: New Constructs of Mind and Reality.* Rochester, Vt.: Park Street Press, 2002.

———. *Evolution's End.* San Francisco: HarperSanFrancisco, 1991.

———. *Exploring the Crack in the Cosmic Egg.* New York: Simon and Schuster, 1973.

————. "Nurturing Intelligence: The Other Side of Nutrition." Address to the Oxford University, World Health Organization, and McCarrison Medical Society Conference on Nutrition and Childbirth, Oxford University, Oxford, England, 1982. See: *Nutrition and Health* 1 (1983): 143–52. London: A. B. Academic Publishers.

Rosenberg, Marshall B. *Nonviolent Communication: A Language of Compassion.* Del Mar, Calif.: PuddleDancer Press, 1999.

West, John Anthony. *Serpent in the Sky: The High Wisdom of Ancient Egypt.* New York: Harper and Row, 1979.

White, Leslie. *A Science of Culture: A Study of Man and Civilization.* New York: Noonday Press, 1969.

INDEX

BOOKS OF RELATED INTEREST

THE DEATH OF RELIGION AND
THE REBIRTH OF SPIRIT
A Return to the Intelligence of the Heart
by Joseph Chilton Pearce

THE CRACK IN THE COSMIC EGG
New Constructs of Mind and Reality
by Joseph Chilton Pearce

SPIRITUAL INITIATION AND THE
BREAKTHROUGH OF CONSCIOUSNESS
The Bond of Power
by Joseph Chilton Pearce

FROM MAGICAL CHILD TO MAGICAL TEEN
A Guide to Adolescent Development
by Joseph Chilton Pearce

A NEW SCIENCE OF LIFE
The Hypothesis of Morphic Resonance
by Rupert Sheldrake

SHATTERING THE MYTHS OF DARWINISM
by Richard Milton

STALKING THE WILD PENDULUM
On the Mechanics of Consciousness
by Itzhak Bentov

THE SPIRITUAL SCIENCE OF THE STARS
A Guide to the Architecture of the Spirit
by Pete Stewart

Inner Traditions • Bear & Company
P.O. Box 388
Rochester, VT 05767
1-800-246-8648
www.InnerTraditions.com

Or contact your local bookseller